Perceiving Geometry
*Geometrical Illusions Explained by
Natural Scene Statistics*

Perceiving Geometry
*Geometrical Illusions Explained by
Natural Scene Statistics*

by

Catherine Q. Howe and Dale Purves

Library of Congress Cataloging-in-Publication Data

Perceiving Geometry: Geometrical Illusions Explained by Natural Scene Statistics /
by Catherine Q. Howe, Dale Purves

A C.I.P. Catalogue record for this book is available
from the Library of Congress.

ISBN-10: 0-387-25487-0   (alk paper)   e-ISBN-10: 0-387-25488-9
ISBN-13: 978-0387-25487-6   e-ISBN-13: 978-0387-25488-3
© 2005 Springer Science+Business Media, Inc.

All rights reserved. This work may not be translated or copied in whole or in part without the written permission of the publisher (Springer Science+Business Media, Inc., 233 Spring Street, New York, NY 10013, USA), except for brief excerpts in connection with reviews or scholarly analysis. Use in connection with any form of information storage and retrieval, electronic adaptation, computer software, or by similar or dissimilar methodology now know or hereafter developed is forbidden.
The use in this publication of trade names, trademarks, service marks and similar terms, even if they are not identified as such, is not to be taken as an expression of opinion as to whether or not they are subject to proprietary rights.
Printed in the United States of America.

9 8 7 6 5 4 3 2 1   SPIN 11385059

springeronline.com

Contents

Acknowledgment vii

1. Introduction 01
2. The Geometry of Natural Scenes 15
3. Line Length 25
4. Angles 37
5. Size 47
6. Distance 63
7. The Müller-Lyer Illusion 73
8. The Poggendorff Illusion 85
9. Implications 97

References 107

Glossary 115

Index 125

Acknowledgment

We are grateful to our colleagues Shuro Nundy and Zhiyong Yang for their collaboration in some of the work discussed in the book, and for much useful advice and discussion. Helpful criticism and advice was also provided by Jim Voyvodic.

Chapter 1

Introduction

Understanding vision—whether from a neurobiological, psychological or philosophical perspective—represents a daunting challenge that has been pursued for millennia. During at least the last few centuries, natural philosophers, and more recently vision scientists, have recognized that a fundamental problem in biological vision—and indeed a fundamental problem in perception generally—is that the sources underlying visual stimuli are unknowable in any direct sense.

The reason for this quandary is the inherent ambiguity of the stimuli that impinge on sensory receptors. In the case of vision, the light that reaches the eye from any scene conflates the contributions of reflectance, illumination and transmittance, as well as a host of subsidiary factors that affect these primary physical parameters (Figure 1.1A). Even more important with respect to the topic under consideration here, spatial properties such as the size, distance and orientation of physical objects are also conflated in light stimuli (Figure 1.1B). As a result, the provenance of light reaching the eye at any moment—and therefore the significance of the stimulus for visually guided behavior—is profoundly uncertain. In more formal terms, this quandary is referred to as the *inverse optics problem*.

These basic facts about the relationship of the real world and the information conveyed by light reaching the retina present a difficult problem. Successful behavior in a complex and potentially hostile environment clearly depends on responding appropriately to the physical *sources* of visual stimuli rather than to the physical characteristics of stimuli as such (which as indicated in Figure 1.1, are of uncertain significance). If the retinal images generated by light stimuli cannot specify the underlying reality an observer must deal with, how then does the visual system produce behavior that is generally successful?

EXPLORING VISION IN TERMS OF PERCEIVED GEOMETRY

In the chapters that follow, we consider the evidence that, with respect to space, the human visual system solves this problem by incorporating past human experience of

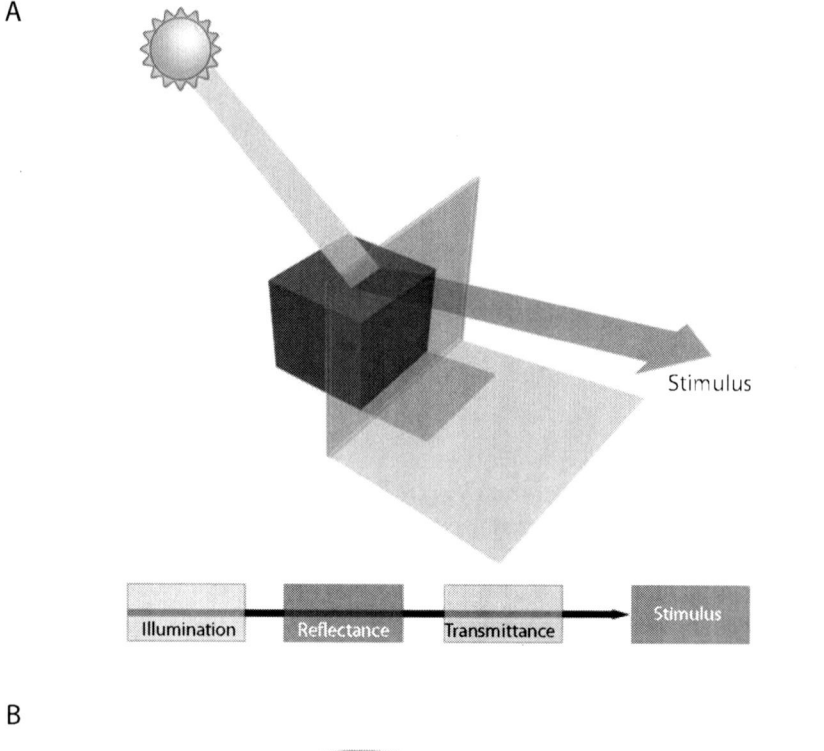

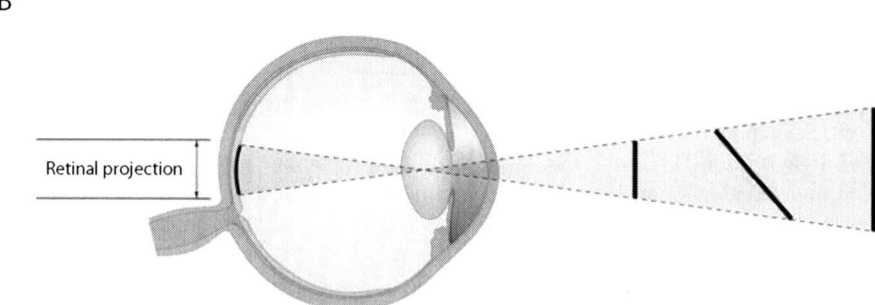

Figure 1.1 The fundamental problem in biological vision is the necessarily uncertain relationship between the information in the images that fall on the retina and their real-world sources. A) Conflation of the factors that determine the amount and spectral quality of light falling on the retina. Illumination depends on the properties of a source like the sun; the reflectance of objects depends on their physical composition; and transmittance depends on the amount and quality of the atmosphere intervening between an object and the observer (as well as between the source of illumination and the illuminated objects). These basic factors that together determine the luminance and spectral distribution of any stimulus at the eye cannot be disentangled by analysis of the retinal image. B) The problem is much the same in the perception of geometry, since the spatial properties of three-dimensional objects are also conflated when light arising from them is projected onto a plane. The diagram shows that the same retinal projection can be generated by objects of different sizes at different distances from the observer, and in different orientations. Again, there is no logical way to disentangle these factors by analysis of the retinal image. (After Purves and Lotto, 2003)

Introduction

with what projected patterns of light on the retina have typically corresponded to in the real world. Indeed, we consider this not simply an adjunct to vision, but the fundamental scheme that determines spatial vision. This empirical strategy explains many otherwise puzzling aspects of what we see, and these explanations—whether in qualitative or quantitative terms—provide the best indication to date of how human perceptions of the geometrical aspects of the world are actually generated.

The crux of the argument is that the link between stimuli and percepts with respect to visual space—i.e., the way we experience size, distance and orientation—can only be understood in a statistical framework in which the perceptions generated by light patterns projected onto the retina are determined by the probability distributions of the possible sources of those projections. This framework can rationalize many otherwise puzzling discrepancies between visual percepts and the physical parameters of visual stimuli. These discrepancies—often presented in the form of "geometrical illusions" (see Figure 1.2)—have long presented a challenge to anyone interested in the nature of vision, and attempts to understand them can be traced back several centuries or more.

Given the inherent ambiguity of retinal images, the biological rationale for seeing the geometrical aspects of retinal stimuli in terms of the probability distributions of their possible sources is not difficult to understand. Much to the advantage of the observer, this visual strategy contends with the problem of stimulus ambiguity by taking

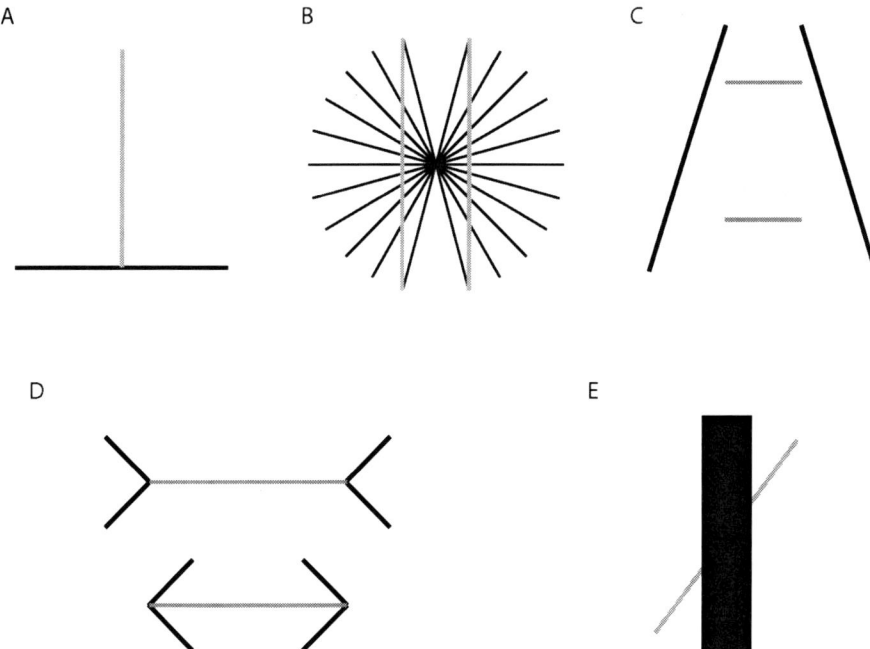

Figure 1.2 Examples of some much-studied geometrical illusions. A) The T-illusion. B) The Hering illusion. C) The Ponzo illusion. D) The Müller-Lyer illusion. E) The Poggendorff illusion.

advantage of eons of trial and error in human experience. In this way we, and presumably all other animals with sophisticated vision, ensure that visually guided responses will usually deal successfully with objects and conditions that are unknowable by any direct means. A consequence of generating percepts on a statistical basis, however, is that what observers actually see does not always correspond to the physical characteristics of the stimulus or the particular physical conditions that generated the stimulus; rather, what is seen corresponds to the empirical significance of the stimulus, i.e., what it has typically meant for visually guided behavior.

Here we examine the validity of this general idea, using as examples a series of classical geometrical stimuli and the "illusions" they generate (the quotation marks are to suggest that this is not a particularly apt word since, in the present framework, *all* visual percepts are equally constructed from the statistical information acquired through experience). The approach in each instance is to examine the statistical relationship between the relevant retinal images and their real-world sources, asking whether the percepts reported by human subjects accord with the predictions made on the basis of the statistics derived from a database of natural scenes. The database includes measurements of the distance and direction of the physical sources of each point in the images, and is effectively a proxy for the accumulated visual experience of our species. As such, the database (see Chapter 2) can be used to reveal the statistical regularities between retinal images and real-world sources that must have determined the evolution of human vision.

Of course, a variety of other visual perceptual qualities can be (and have been) explored in this framework, including brightness/lightness, color and some aspects of motion (reviewed in Purves and Lotto, 2003). We have chosen to focus here on perceived spatial relationships because understanding the way we see geometry is intrinsically interesting and much debated; a further reason is that, from a technical perspective, the probability distributions of the possible sources of geometrical stimuli can be derived directly by an analysis of range images of natural scenes. The ability to determine the statistical relationship between geometrical projections and their sources is an enormous advantage, and meeting this goal is much more difficult (although certainly not impossible) with respect to other stimulus characteristics.

Of course, the fact that geometrical stimuli and the perceptual anomalies they elicit have challenged the imaginations of so many thinkers in a variety disciplines over so many years is strong motivation as well. In the end, however, the point of the work summarized here is simply to understand how and why we perceive visual space and the geometry of the objects therein the way we do.

EXAMPLES OF WELL-KNOWN GEOMETRICAL ILLUSIONS

Numerous observers have pointed out that measurements made with rulers or protractors of a variety of simple visual stimuli are often at odds with the perceptions they elicit, frequently in a most engaging way. Constructing stimuli that produce a particularly intriguing geometrical illusion was something of a cottage industry in the 19th and early 20th centuries; although some of these demonstrations were described by the preeminent vision scientists of the time, a good geometrical illusion has provided

Introduction

eponymous immortality to a number of investigators whose names would otherwise not be known today (reviewed in Luckiesh, 1922; Coren and Girgus, 1978; Rock, 1995; Robinson, 1998—the book by Robinson provides an especially detailed and authoritative history of this field).

Some of the best-known geometrical illusions—and the ones whose etiology has been most hotly debated—are illustrated in Figure 1.2. Perhaps the simplest of these is the "vertical-horizontal" or "T-illusion" attributed to Johann Joseph Oppel (1855), in which the vertical line appears longer than the horizontal line, despite the fact that they are of equal length (Figure 1.2A). Oppel is also credited with having coined the term "geometrical illusion" (Robinson, 1998). A more elaborate example attributed to Ewald Hering (1861) shows two parallel lines (indicated in red) that appear bowed away from each other when presented on a background of converging lines (Figure 1.2B). In the Ponzo illusion (created by Mario Ponzo in 1928) the upper horizontal line appears longer than the lower one, despite the fact that they are again identical (Figure 1.2C). In the more complex Müller-Lyer illusion (created by Franz Müller-Lyer in 1889), the line terminated by arrow tails looks longer than the same line terminated by arrowheads (Figure 1.2D). The final example, also dating from the 19th C., was devised by Johann Poggendorff (Figure 1.2E). In this stimulus, the continuation of a line interrupted by a bar appears to be displaced vertically, even though the two line segments are actually collinear. Many other instances of the discrepancies between the objective geometrical features of a stimulus and the percept it gives rise to can be found in various popular books on illusions (see Sackel, 2000 for an especially good example of this genre). Most of these, however, are variations on the basic themes illustrated in Figure 1.2.

It would be a mistake to conclude that, because the stimuli in Figure 1.2 are simple geometrical figures that bear little resemblance to real-world objects, these discrepancies between stimulus and percept are not significant for behavior in natural environments. Figure 1.3 shows that, even when such stimuli are presented in more realistic settings, the perceptual effects elicited by the simpler versions in Figure 1.2 persist.

EXPLANATIONS BASED ON "MISINTERPRETATION" OF RETINAL IMAGES

Despite a great deal of work and much speculation, there is no consensus about the basis of these geometrical illusions, much less about why, in more general terms, we see geometrical forms in the peculiar way we do.

Some of the early attempts to rationalize these phenomena, such as explanations based on asymmetries in the anatomy of eye or the ergonomics of eye movements (see Robinson, 1998, p.138 ff.), were clearly off the mark given modern evidence about the workings of the visual system. However, ever since Hermann von Helmholtz suggested in the latter half of the 19th C. that the visual brain makes "unconscious inferences" about the underlying nature of scenes, most theories have assumed that observers depend to some degree on information from past experience to help "interpret" retinal images. Thus, how this information is used (or misused) has been thought to in some way

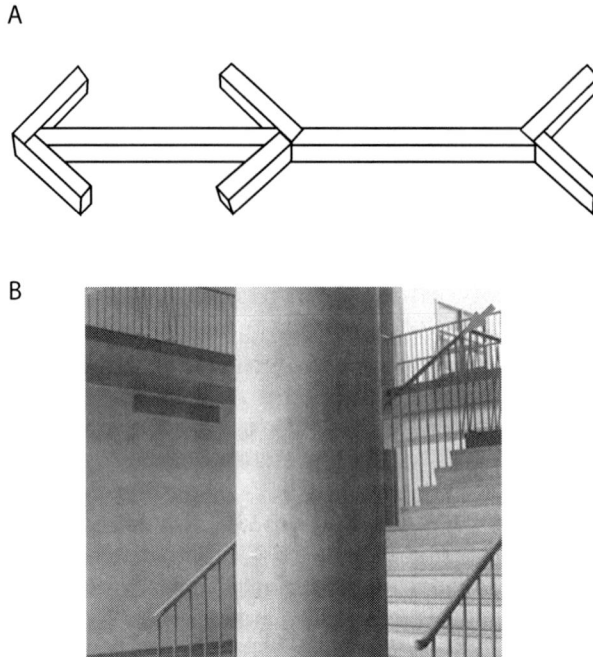

Figure 1.3 The Müller-Lyer (A) and the Poggendorff (B) illusions presented in more natural forms. The effects apparent in Figure 1.2 persist, ruling out the idea that these perceptual phenomena are only evident "in the laboratory". (A is after Fisher, 1970; the photograph in B was taken in the lobby of our building; the arrows indicate the interrupted collinear object, as in Figure 1.2E.)

account for these "misperceptions" of the actual dimensions and spatial arrangement of stimuli whose physical sources are not defined in the image. As Helmholtz put it, "In the case of uncertain perception, our judgment is apt to be led astray by other causes that affect it" (Helmholtz, 1924, vol. III, p. 188). (Oddly, Helmholtz, a brilliant polymath whose thinking remains deeply influential today in both vision science and physics, attributed many of these effects to eye movements.)

Modern investigators have for the most part imagined these "other causes" to be a variety of hidden assumptions (sometimes called "heuristics") made by observers as a result of their experience with objects in the world. For example, Armand Thiéry suggested in the late 19th C. that an observer's interpretation of 2-D retinal images as perspective projections of particular 3-D objects might be the basis for at least some geometrical illusions (Thiéry, 1896). Following this general line of reasoning, Richard Gregory argued in the 1960s that the Müller-Lyer illusion (see Figure 1.2D) is a result of seeing the arrow-tails and arrowheads figures as "concave" and "convex" corners, respectively, in the 3-D world; in this scenario, the anomalous percept is taken to be an unconscious result of previous experience with the different distances from the eye implied by such real-world corners (Gregory, 1963; 1968). (Intuitively it seems that convex corners will have been generally closer than concave ones; see, however, Chapter 7.) Gregory (1966) also used this type of explanation to rationalize

Introduction

the apparently different length of the two lines in the Ponzo illusion (see Figure 1.2C). The so-called "inducing lines" in the Ponzo stimulus converge, much as the appearance of railroad tracks extending into the distance. As a result of such experience, so the theory goes, observers would unconsciously "assume" that the upper horizontal line is further away than the lower one. Given this assumption, the upper line would then have to be a physically longer object, and its projection might therefore be expected to *look* longer, as it does. In a similar vein, Barbara Gillam (1998) has suggested more recently that several other geometrical illusions are based on the observer's familiarity with the scale and size of objects, and the way these parameters are typically affected by the perspective generated in projected images.

A problem with explanations of this general sort is that some geometrical illusions persist even when the stimulus entails little or no explicit information about depth. The Müller-Lyer illusion, for instance, is seen even if the ends of the two lines are terminated by circles or squares instead of arrowheads and arrow tails (Figure 1.4A). Conversely, the presence of depth information does not always guarantee an illusory effect. For example, the Ponzo illusion is much diminished when the two lines to be compared are presented as vertical attachments to one of the inducing lines (Rock, 1995) (Figure 1.4B).

These confounding facts led some psychologists to endorse yet another explanation for geometrical illusions, known as the theory of "contrast and confluence" (Obonai, 1954; Rock 1995; these concepts were initially described by Helmholtz [1924, vol. III, pp. 237–240]). In this interpretation, observers are imagined to perceive an object's properties based on a comparison with other nearby features. In the case of "contrast", any differences observed between the object and the context would, in this conception, tend to be exaggerated by the observer; in the case of "confluence", the properties of the context are taken to be assimilated into the

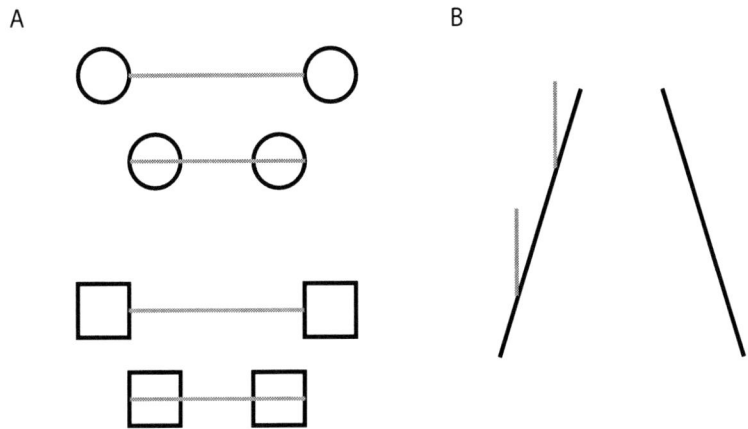

Figure 1.4 Examples that confound intuitive explanations of some classical geometrical illusions. A) Persistence of the Müller-Lyer illusion in the absence of possible differences in depth implied by arrowheads and arrow tails. B) Diminution of the Ponzo effect by an altered arrangement that nonetheless maintains the depth information conveyed by the converging lines.

perceived qualities of the object. For example, the lower line in the Ponzo illusion would, in this interpretation, tend to look smaller in contrast to the relatively large empty space at either end of it, whereas the upper line would tend to look larger in contrast to the relatively small space at its ends. The same general idea has also been proposed to explain the size contrast and assimilation effects taken up in Chapter 5. The problem with this sort of account is its limited ability to explain the variety of geometrical illusions that have been described; for example, this explanation cannot account for most of the standard effects illustrated in Figure 1.2, or the variants in Figure 1.4.

In summary, the theme of these several explanations of geometrical illusions is that experience modulates the appearance of a world that would, under more usual circumstances, be accurately represented by the "sensations" generated by "low-level" visual processing. A corollary is that observers tolerate these discrepancies as a "debt which the visual system pays for other [unspecified] advantages in the process of seeing" (Robinson, 1998, p. 253).

ECOLOGICALLY BASED EXPLANATIONS

A more radical approach to rationalizing the perception of geometry is James Gibson's theory of "ecological optics" (1966, 1979). In effect, Gibson argued that the significance of visual stimuli, including those that give rise to geometrical illusions, is only ambiguous when presented in the stylized form used in most psychophysical experiments. His point was that natural stimuli contain ample information to allow unambiguous, and, in Gibson's terminology, "direct" perception of the real 3-D world. In this way of thinking about the problem, geometrical illusions are largely the result of impoverished stimuli. Gibson thus identified a series of "higher-order" variables such as texture gradients in the structure of what he called the "ambient optical array" (i.e., the pattern of the light that comes to an actively exploring observer in terrestrial settings), arguing that a direct apprehension of these common features of the real world is the basis of human spatial percepts.

Despite a number of important and original insights, Gibson's concept of vision does not really explain how the information in the "ambient optical array" determines what observers see. Nor does it explain the persistence of many geometrical illusions in fully natural viewing circumstances (Higashiyama, 1996; Dixon and Proffitt, 2002; see also Figure 1.3). More importantly, however, it does not address the fundamental problem of stimulus ambiguity outlined earlier. Whereas the wealth of information in any natural scene is certainly influential in determining what an observer sees, the richness of detail in a natural or any other complex scene does not alter the problem of relating the uncertain significance of the elements in a retinal image to the real world in which the observer must behave (see Figure 1.1). Although retinal images generated in natural environments certainly contain more components than a stylized stimulus constructed in the laboratory, these elements are just as ambiguous as the components of any artificial stimulus. The fact that observers routinely generate visual perceptions that enable them to deal successfully with the environment is not evidence that "naturally" generated retinal images lack ambiguity, but rather that evolution and development have found a way to solve the basic problem in vision illustrated in Figure 1.1.

FEATURE DETECTION AND RULE-BASED SCHEMES OF VISION

The prolonged debate about the basis of geometrical illusions notwithstanding, the psychological inquiries of Helmholtz, Gibson and many others into how the visual system generates useful percepts in the face of stimulus ambiguity have been largely overshadowed since the 1950s by rapid advances in the study of visual physiology, and by the advent of computer vision.

The discovery by David Hubel and Torsten Wiesel (1959, 1962, 1968, 1974) of cells in visual cortex that selectively respond to edges at different orientations, and the subsequent discovery of the selective responses (i.e., the "tuning") of cortical neurons to a variety of other retinal image features such as direction of motion, speed, spatial frequency and color (see, for example Hubel, 1982), shifted the focus of vision research toward the question of how retinal stimuli are "represented" at various levels of the visual processing pathway. Although such studies do not explicitly address the neural basis of perception (much less geometrical illusions), an implicit assumption has been that the response properties of higher-order cells in the primary and extra-striate visual cortices revealed in this way would ultimately provide a physiological explanation of the relevant percepts (i.e., that the responses of visual neurons at some level correspond directly to conscious percepts). At the same time, enormous advances in the theory and practice of computer vision promised the means of modeling how feature-detecting algorithms might actually work (see, for example, Marr, 1982).

Given what has already been said about the inverse optics problem and its fundamental place in any concept of how vision works, it should be apparent that visual percepts are unlikely to be explained by understanding how the features of the retinal image are encoded by visual neurons (although the properties of visual neurons obviously will be important in whatever understanding of visual perception eventually emerges). If, by their nature, retinal images do not allow a unique interpretation of the underlying 3-D world that observers must respond to, any strategy based on an analysis of the retinal image as such seems doomed to failure. Nevertheless, an unstated assumption of both physiologists and computer scientists interested in vision over the last few decades has often been that by combining knowledge of visual physiology and Helmholtz's concept of unconscious inference, a representation of the world can somehow be constructed from encoded retinal image features according to a set of logical rules instantiated in visual processing circuitry (i.e., the hidden assumptions about the physical world or the heuristics referred to earlier).

The problem for any rule-based scheme of vision can be appreciated by considering a computer program that seeks to play out the all logical sequences that would determine the best response to a given board position in chess. Completion of this task in a finite time defies even the most powerful computer imaginable because of the astronomical number of possible sequences entailed in playing out all possible chess games (estimated to be on the order of 10^{120}). Since the skein of possible relationships between all the points in a retinal image and their possible sources in the world is far more complex than the possible moves in chess, linking the components of retinal images and their generative sources to enable rapid behavioral responses according to a set of logical rules is simply not a viable strategy.

A WHOLLY PROBABILISTIC FRAMEWORK OF VISION

A more plausible scenario for the way the visual system contends with inherently ambiguous stimuli is to operate probabilistically, generating percepts entirely determined by the past success or failure of visually guided behavior. By gradually accumulating empirical evidence about the linkage between images and their sources in this way, and by continually adjusting both percepts and behaviors according to this growing body of experience, the evolving visual system would eventually be able to routinely generate successful visually guided responses to retinal images. Although in this highly evolved state percepts and visually guided behavior would give the appearance of being generated according to logical rules, they would in fact be determined in a purely statistical way that simply reflected the accumulated influence of all past experience on the present structure of visual processing circuitry.

In this way of thinking about vision, the spatial perceptions elicited by the geometrical (or any other) aspects of visual stimuli are determined by the probability distributions of all the possible sources of the relevant retinal images. Indeed, there appears to be no other way, in principle, to contend with the uncertain relationship of projected images and their sources.

TESTING THIS IDEA

If the uncertain provenance of stimuli is indeed resolved entirely on the basis of image-source statistics, then it should be possible to predict how observers will perceive the qualities of any visual stimulus based on the probability distribution of the real-world sources of a given retinal projection. Indeed, evaluating the scope and accuracy of such predictions would be the best way to test the merits of the proposition that vision is wholly predicated on the statistical relationship between images and their physical sources.

Assessing the probabilistic relationship between retinal images and their sources is, of course, an extraordinarily difficult task that would ultimately require analysis of a very large database of natural scenes with complete information about the distance, luminance and spectral characteristics of all the elements in the images. Nevertheless, as noted earlier, this aim can be pursued straightforwardly in the relatively limited domain of spatial vision. The approach we used to gather the evidence described in the chapters that follow was to acquire a database of natural images that included information about the location in 3-D space of every element (pixel) in a series of representative visual

Introduction 11

scenes, as described in Chapter 2. By statistical analysis of the geometrical relationships between images and sources in the database, we could compare the percepts predicted on this basis with the perceptions of form, distance and direction that observers actually see.

OTHER EMPIRICAL INFORMATION PERTINENT TO THE PERCEPTION OF VISUAL SPACE

There are, of course, many categories of empirical information in addition to the distance and direction of each point in a scene that are pertinent to perceiving the spatial arrangement of objects. The sense of three-dimensionality in Figure 1.5, for instance, arises from a wealth of mutually consistent information about the probable physical sources of the objects in the image. Examples are perspective, occlusion, attenuation of light by the atmosphere, shadowing, the changing texture of the background and so on. Thus the perception of geometrical relationships depends on much more than experience with size, distance and direction as such. This body of additional information is usually described in textbooks under the rubric of "monocular cues to depth" (in contrast to the specific depth information that arises from slight disparities between the left and the right retinal images, which is referred to as stereoscopic or binocular information).

Perhaps the most obvious source of monocular information pertinent to spatial arrangement is *occlusion*: when part of one object is obscured by another, it will always have been the case that the obstructing object is closer to the observer than the obstructed object. Another source of information about depth is *aerial perspective*: because Earth has a substantial atmosphere, the further away objects are from the observer, the more the interposed matter, which makes objects look fainter and fuzzier as a function of distance. Moreover, because the atmosphere absorbs more long than short wavelength light (the interposed medium is effectively sky), distant objects also look bluer compared to their appearance nearby, as landscape artists have long recognized (Minnaert, 1937). Shading and shadow also provide a body of detailed information about spatial arrangement. A final general category is motion parallax. When the position of the observer changes (by moving the head or body), the position of the background with respect to an object in the foreground changes more for nearby objects than distant ones.

It seems obvious that this sort of empirical information about depth must be acquired. Observers discover through experience that more distant objects are more often occluded, smaller in appearance, fainter, fuzzier and bluer, and that they tend to change position less with respect to the background when the head is moved. It would be a mistake to assume, however, that any information pertinent to the way we see visual space is incorporated into the nervous system by learning in the ordinary sense of this word, i.e., by the influence of experience during an individual's lifetime. Although learning during ontogeny obviously occurs and is demonstrably important in many aspects of vision (see Chapter 13 in Purves and Lichtman, 1985 for a review), the statistical relationship between images and sources pertinent to the most fundamental

12   **Perceiving Geometry: Geometrical Illusions Explained by Natural Scene Statistics**

Figure 1.5   The sense of depth in a complex natural image such as this is the result of mutually consistent information pertinent to the spatial relationships of objects in the scene, including occlusion, perspective, texture gradients, atmospheric effects ("aerial perspective"), shading and shadow and other factors.

aspects of seeing have, as already implied, been built into the visual system by eons of phylogenetic experience, as well as by whatever individual experience adds during a person's lifetime.

The relative contributions of ontogeny and phylogeny to the empirical information reflected in the circuitry of the visual system will always be a somewhat murky issue (as will any aspect of such "nature-nurture" questions). Nevertheless, it should be clear that *many* empirical factors contribute to the perception of geometry, and that a reasonably complete understanding of even this relatively simple aspect of visual perception will need to take account of the statistical influence of at least the most basic of these factors.

Introduction

SUMMARY

In addition to the conflation of the physical parameters that determine the quality and quantity of light that reaches the eye from the objects in any scene (the primary factors being illumination, reflectance and transmittance), the parameters that define the location and arrangement of the sources of light—the size, distance and orientation of objects—are inextricably intertwined in the retinal image. As a result, the spatial relationships of the sources of visual stimuli are always uncertain. Subsequent chapters review the evidence that the human visual system contends with this problem of ambiguity by generating geometrical percepts according to the probability distributions of the possible real-world sources of retinal stimuli. The signature of this solution with respect to the perception of geometry in a visual scene is a wealth of subtle (and sometimes not so subtle) discrepancies between spatial percepts and the metrics of the stimuli that generate them (i.e., geometrical illusions). The argument that follows is that the existence of these discrepancies and their extraordinarily complex phenomenology can only be understood in terms of an entirely probablilistic framework of vision.

Chapter 2

The Geometry of Natural Scenes

Given this overall framework for understanding the genesis of geometrical illusions and spatial percepts generally, information about the 3-D structure of the world is obviously essential. If the goal is to relate the geometry of retinal projections to their possible sources, a database is needed that fully details the spatial arrangement of objects in the sorts of scenes that humans have typically witnessed. Given this information, it should then be possible to test whether the peculiar perceptions elicited by the sorts of geometrical stimuli described in Chapter 1—and indeed any geometrical percept—can be rationalized on the basis of the statistical relationship between the two dimensional images projected onto the retina and their three dimensional sources in the physical world.

Although acquiring detailed spatial information about the structure of the real world would have been difficult to imagine not too many years ago, rapid advances in technology have made at least one aspect of this problem straightforward. In the construction industry, the progress of building projects and their geometrical conformance with plans is now routinely monitored using a method called laser range scanning. The technique provides accurate measurements of the distances of all the points (pixels) in a digitized scene from the origin of the scanner's laser beam.

Happily, this device is well suited to acquiring the database needed to explore the perception of visual space. Since the height of the scanner can be set at the average eye-level of a human observer (as it was routinely in collecting the data we used), and since there is no laser return from surfaces obscured by intervening objects (thus taking occlusion into account), the information acquired in this way provides a good first approximation of the spatial characteristics of the scenes humans typically see.

ACQUIRING A DATABASE OF NATURAL SCENE GEOMETRY

The natural scene database for the various analyses described in the chapters that follow were acquired with a high-precision scanner (Figure 2.1) that combined the sort of laser range-finder just described with a channel capable of sensing light intensity, thus providing digitized images with luminance and color information for each pixel

16 Perceiving Geometry: Geometrical Illusions Explained by Natural Scene Statistics

Figure 2.1 Range scanning apparatus used to determine the physical geometry of scenes. The way this device works is quite remarkable (see Besl, 1988). A scanning mirror inside the rotating optical head directs a laser beam over a precise angular pattern. The beam of collimated infrared light is pulsed periodically from the laser, such that each point (pixel) in the scene is evaluated sequentially. The signal reflected back from object surfaces is detected by a photodiode, which in turn produces an electrical "receiver signal". The tiny interval of time between the transmitted pulse and signal returned is determined by a quartz-stabilized clock; based on the speed of light, the distance from the scanner to each surface point is then calculated by a microcomputer. The distances determined in this way are accurate to within a few millimeters. The spatial resolution (angle step-width) of this process is selectable between $0.072°$ to $0.36°$ in both elevation and azimuth. For practical reasons, we chose $0.144°$; at this resolution it takes approximately 3 minutes to scan an image that encompasses the standard "field of view" of the scanner, an acquisition time that works well in the field.

in the scene (i.e., the information that generates ordinary photographs), as well as information about distance and direction. The performance of the range-finding aspect of this apparatus is effective from a minimum distance of about 2m to approximately 300m, with an accuracy of ± 25mm and an angular resolution of up to $\sim 0.072°$ (for reasons explained in the legend of Figure 2.1, we used a lower resolution than the maximum capability of the device).

The scanner was mounted on a surveyor's tripod such that the origin of the laser beam was always at a height of 165cm (the average height of the adult human viewpoint), and the apparatus leveled in the horizontal plane before acquiring each

The Geometry of Natural Scenes

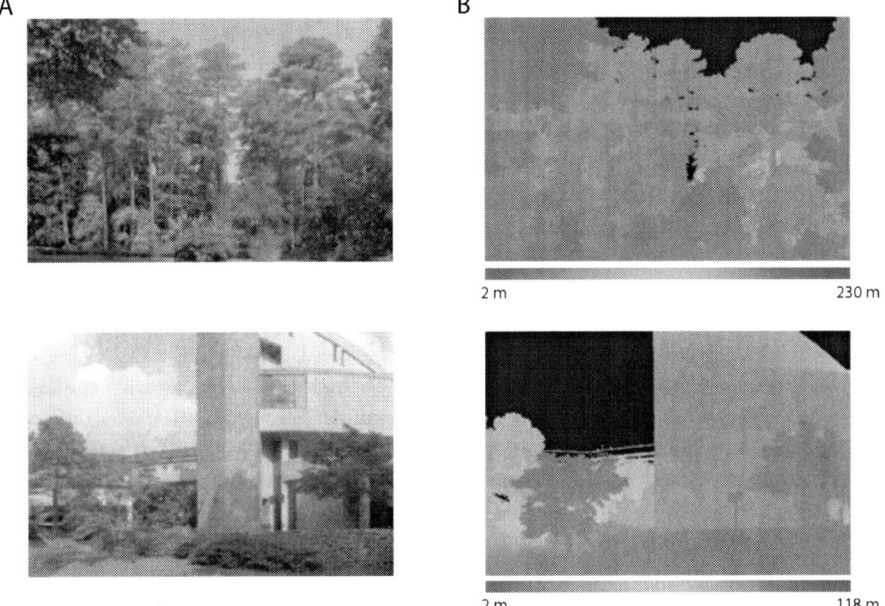

Figure 2.2 Representative images acquired by the laser range scanning apparatus; each of these images covers about 65° horizontally and 45° vertically, and is thus a relatively small portion of the full wide-field scan. (A) Ordinary color images of a wholly natural scene and an outdoor scene that contains human artifacts. (B) The corresponding range images acquired by the laser scanner. The physical distance of each point in the scene from the origin of laser beam is indicated by color-coding; black areas are the points in the scene from which no laser reflection was recorded (the sky); such points were omitted from subsequent analyses.

image. Using this system, we collected a large number of images in different settings on or near the Duke University campus. About a quarter of the images were fully natural scenes (i.e, that contained few if any human artifacts), whereas the rest included buildings and other constructions (of these about two-thirds were taken outdoors and one-third indoors). Examples of a fully natural scene and a typical scene containing human constructions are shown in Figure 2.2. Since the scanner covered 333° horizontally and 80° vertically (the standard field of view of the scanner), these examples show only a fraction of each scanned image.

The large number of laser scanned images acquired in each environment ensured a representative and thematically diverse sampling of the human visual environment. As will be apparent in later chapters, categorizing the images as wholly natural versus those containing human constructions was useful in addressing some of the specific debates about geometrical illusions that have persisted over the years.

SOME LIMITATIONS OF THIS SORT OF DATABASE

It should be clear that the information acquired in this way about the physical distance and direction of all the points on surfaces in the visual environment is an incomplete indication of all aspects of the spatial arrangement of objects in a scene. Thus a number of relevant limitations need to be borne in mind.

The first of these is the restricted range of distances over which the laser scanner operates (~2–300m) and relatively low angular resolution that we used (see above). Since human beings obviously see things that are closer than 2m and more distant than 300m, and can readily resolve points more closely spaced than the scanner resolution that we used (normal human vision resolves points as close as ~0.005°), the information obtained in this way is only an approximation of the range and resolution of the scenes that humans routinely see. Using information from this relatively restricted range and resolution requires the assumption that the statistics obtained from the database are more or less representative of the additional information that would have been gleaned for the more highly resolved nearer and further objects that we see and interact with on a daily basis. Moreover, we avoided placing the scanner in front of nearby objects, which would have blocked out the rest of the scene during the acquisition of an image. As a result, the points in the database are somewhat biased toward greater distances. These necessary assumptions about range and resolution seem reasonable, but are nonetheless deficiencies.

Another deficiency is that the database obviously does not represent the variety of landscapes and spatial arrangements found in natural environments worldwide; all the scenes in the database were acquired in a particular locale (on or near the Duke University campus), in a particular season (summer), at a particular time of day (full daylight) and in limited weather conditions (days on which the equipment was not likely to get rained on).

A further limitation is that other visual information pertinent to the perception of scene geometry such as stereoscopic disparity (the basis of binocular depth sensations), spectral distribution (the basis of color vision) and motion parallax was not included in the database. These additional data, which would be more difficult or impossible to obtain with present technology, will eventually be needed to fully assess the probabilistic relationship between the photometric patterns in the image plane and the generative physical geometry (see the section on "Other empirical information pertinent to the perception of visual space" in Chapter 1).

A final concern is that humans (or other visual animals) do not simply observe the world in the systematic fashion of the laser scanner, but fixate on objects and regions of objects that contain information that is particularly pertinent to subsequent behavior. As the Russian physiologist Alfred Yarbus first showed some 50 years ago, human observers are highly biased in the time they devote to viewing different components of scenes (reviewed in Yarbus, 1957; see also Parkhurst and Niebur, 2003). Since the scanner samples all portions of a scene uniformly, and since we had no principled way to incorporate these human biases post hoc, this deficiency is also inherent in our analysis.

USING INFORMATION IN THE DATABASE

The basic challenge in vision outlined in Chapter 1 is understanding how the visual system manages to generate biologically useful percepts from retinal stimuli that are

The Geometry of Natural Scenes

inherently ambiguous. The broad hypothesis about the way observers meet this challenge is that percepts are determined by the probability distributions of the possible sources of visual stimuli, and that this accumulated empirical information has been instantiated in the structure of visual circuitry through trial and error over both evolutionary and developmental time.

The specific question this concept raises with respect to spatial vision is thus whether the geometry we actually see is in all cases accurately predicted by the statistical relationship between images and their physical sources. Such evidence would validate the general proposition that humans and other highly visual animals solve the inverse optics problem by means of this wholly probabilistic strategy; at the same time this evidence should explain the full range of anomalies in the perception of geometry that have puzzled thinkers over the years (see Chapter 1).

The general approach we have taken to testing these ideas and predictions about spatial vision and the wealth of geometrical illusions that have now been described is to sample the images in the database with geometrical templates configured in the same form as a stimulus pattern of interest. The range information can then be used to determine the corresponding physical location of the relevant points in 3-D space, in this way relating the 2-D image to its 3-D sources. By sampling a large set of points pertinent to the stimulus of interest in many different images, the probabilistic relationship between the geometry of the stimulus in the image plane and the spatial properties of its possible physical sources can then be determined, at least within the sorts of limitations mentioned earlier. Using these statistical relationships, we could then ask whether the perception of any given geometrical stimulus is, in fact, accurately predicted on this basis.

RELATING RETINAL IMAGES TO THE PHYSICAL WORLD

The key issue in thinking about the merits of this general approach to understanding perception is how the statistical characteristics of images are related to the structure of their generative sources in the real world, and how this relationship is used to biological advantage.

It should be intuitively obvious that regularities in the structure of the world must be represented in projected images, and that visual animals would wish to take advantage of this information in generating behavior that contends successfully with the inverse optics problem. What, then, is the nature of this statistical connection, and how could it be used? One way of thinking about this relationship would be to imagine that because regularities of the world are necessarily embedded in the statistical structure of images, exposure to a large number of images *as such* would be sufficient for the visual system to generate appropriate percepts and behavior. On the face of it, this perspective makes sense. After all, the information in the retinal image is all the visual sensory system has to work with in terms of present stimuli. In fact, this conception is badly misleading. Absent behavioral interactions with the physical sources of stimuli, there is no way for a human observer or any other visual animal to make the link between the statistical structure of images and what the images signify in terms of the physical world the observer must contend with. A disembodied visual system (a computer with visual input, for example) would not be able to use a compilation of images per se to understand the nature of its environment, no matter how extensive or how thoroughly analyzed. Thus some form of interaction and feedback is needed to understand how the images are linked to their real-world sources, which is, in the end, the basis for successful visually guided behavior.

Accordingly, the biological significance of image-source relationships can only be acquired by visual observers who are actually behaving in the world that generated the images, thus accumulating information about the relative success or failure of their actions. The range image database and its use here effectively serve as a proxy for this experience. Linking ambiguous spatial information in projected images to the actual location in 3-D space of all the points in the scene by laser range scanning mimics, for all intents and purposes, what observers would have ordinarily learned by acting in the environment, in this way correlating retinal images with the underlying arrangement of physical objects and the pertinent behavioral implications (again within the limitations of the database mentioned earlier).

In short, successful spatial perceptions and actions require linking spatial relationships in projected images to the corresponding physical sources. In a biological context, animals presumably make this linkage by an ongoing empirical tally of the success or failure of behavior in response to the variety of visual stimuli produced by the natural environment on our planet. The agent for this accumulation in the first instance is natural selection operating on the inherited organization of the visual system, and in the second instance mechanisms of neural plasticity that operate on the organization of the visual system over the lifespan of individuals. Both mechanisms lead to an ever increasing sum of empirical information in the brains of visual observers.

OTHER ASPECTS OF IMAGE STATISTICS AND THEIR RELEVANCE TO VISION

Despite the requirement that visual systems somehow incorporate the probabilistic relationship between images and their physical sources to generate appropriate percepts and behavior, the statistical information in images per se has proved quite useful in vision research, particularly in thinking about functional optimization. For example, a good deal of recent work on natural images and their statistical properties (reviewed in Simoncelli and Olshausen, 2001) has been motivated by the plausible assumption that the visual system of humans or other animals would wish to take advantage of the statistical regularities in projected images to encode the information in retinal stimuli with optimal efficiency.

Indeed, the intrinsic statistical properties of natural images are very likely to be key determinants of optimal coding strategies, as many have argued (Barlow, 1961; Atick and Redlich, 1992; Field, 1994; see also Ruderman and Bialek, 1994; Chiao et al., 2000; Schwartz and Simoncelli, 2001; Turiel et al., 2001). At the very least, efficient information transmission must rank high as an evolutionary force that has also contributed to shaping the structure and function of the visual system. In any

The Geometry of Natural Scenes

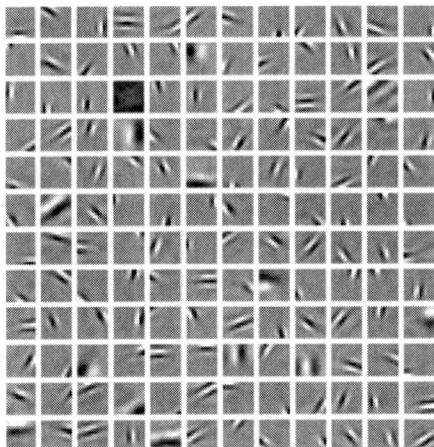

Figure 2.3 Each square in this figure is one of a set of statistically independent components derived from the analysis of a large number of natural image patches (12 × 12 pixels). Each of the original image patches could be reconstructed by a linear superposition of the components illustrated. (From Simoncelli and Olshausen, 2001)

event, the statistical structure of natural scenes has been examined in gray scale images (van Hateren and van der Schaaf, 1998), color images (Wachtler, et al., 2001), stereo images (Hoyer and Hyvärinen, 2000) and motion sequences (Dong and Attick, 1995; van Hateren and Ruderman, 1998). The results of such studies have generally supported the idea that the response properties of at least some classes of visual cortical neurons are influenced by these statistics.

For example, if natural images are parsed to determine their statistically independent components, the components are found to contain the sort of spatially localized and orientated structures shown in Figure 2.3 (Olshausen and Field, 1996, 1997; Bell and Sejnowski 1997; van Hateren and van der Schaaf, 1998). The significance of this finding is that the independent image components are structurally similar to the receptive fields of some neurons in the primary visual cortex (V1). Neuronal receptive fields are generally defined as the area of the overall visual field within which the presentation of a stimulus elicits responses in the neuron under consideration. The receptive fields of neurons in V1 are often elongated in particular orientations. The similarity between the independent components of natural images in Figure 2.3 and the receptive fields of V1 neurons suggests that visual neurons act as filters that form "sparse" representations of natural images, thus minimizing the redundancy of the information transmitted. Observations of this sort thus support the idea that the response properties of visual neurons have been molded by the statistical characteristics of natural images to optimize information transfer.

Despite the usefulness of the intrinsic statistics of images as a means of illuminating efficient visual coding strategies, this perspective does not address what we take to be the fundamental problem in vision, namely contending with the direct unknowability of the physical world by means of the information in projected images.

LINKING IMAGES AND THEIR SOURCES BY EMPIRICAL RANKING

There are a number of ways in which the statistical relationship between images and sources can be used to predict the geometry that people would be expected to see in response to various geometrical stimuli (see, for instance, Nundy et al., 2000; Howe and Purves, 2002; Yang and Purves, 2003). The method used in the work described in the following chapters was chosen because of its biological plausibility and its ability to explain the full range of geometrical phenomena that have been described over the decades (for comparison with a Bayesian approach, see supplement in Howe and Purves, 2005a).

In this approach, referred to as empirical ranking, the different perceptual qualities evoked by any retinal stimulus (e.g., perceived size, orientation, brightness) are determined by the rank of the corresponding stimulus characteristic (e.g., the size, orientation or luminance of the feature of interest in the projected image) within the full range of past human experience. To illustrate this approach, consider the perceived length of a line that has a certain projected length in the retinal image. The perceived length, in this framework, is determined by the percentile rank of the projected length of that line within the entire range of projected lengths that have been experienced by human observers over the course of evolution and individual development. This empirical range can be defined more specifically as a scale determined by the frequency of occurrence of the real-world sources that have given rise to linear projections of various lengths. For instance, if in past human experience 25% of the possible physical sources of lines generated stimuli shorter than or equal to the length of the particular stimulus line in question, the rank of that projected length would be the 25th percentile. If, on the other hand, the length of another line stimulus has a rank of, say, the 30th percentile, then the stimulus at the 25th percentile should appear shorter. In other words, relative perceptual magnitude follows percentile ranking.

Notice that the empirical rank of a stimulus characteristic derived in this way will almost always be different from its rank on a linear scale of that stimulus feature (Figure 2.4). If, for instance, the minimum possible length of projected lines is taken to be 1 and the maximum length 100, then a line 25 units in length would rank at the 25th percentile on a linear scale extending from 1 to 100. The empirical scale of line length, however, takes into account how *frequently* the physical sources of linear stimuli shorter or longer than 25 units have actually been encountered in the accumulated sum of human experience. If the physical sources of the stimulus lines shorter than 25 units have occurred more often than the sources of the lines that are longer, then the empirical ranking of the 25-unit stimulus line would be higher than the 25th percentile, and conversely. In this way, past experience is incorporated in the process that generates visual percepts. A consequence of this process is, of course, discrepancies between the linear metrics that apply to projected images and the subjective "metrics" that characterize perception.

In the chapter that follows, we apply this approach to explain the puzzling variation in the perceived length of a line as a function of its orientation in the projected image.

The Geometry of Natural Scenes

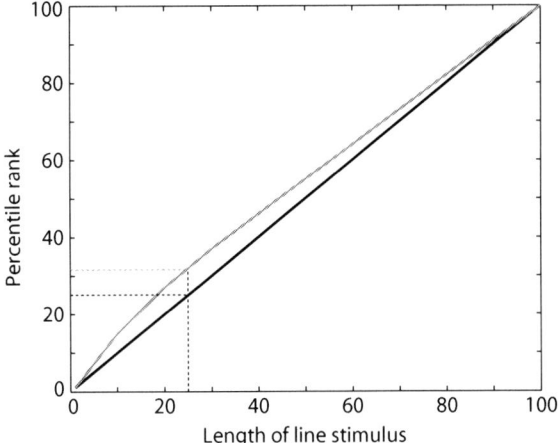

Figure 2.4 A comparison of linear and empirical scales, using stimulus length as a simple example. The black line indicates stimulus length on a linear scale that extends from 1 to 100. The gray curve, in contrast, illustrates a hypothetical empirical scale of stimulus length based on how frequently physical sources that give rise to different projected lengths have occurred in past experience. A line of a given length in the retinal image (e.g., 25 units) has a different rank on these two scales, as indicated by the dotted lines.

The same approach is also used subsequently to explain the variety of other phenomena apparent in perceived geometry.

THE BIOLOGICAL RATIONALE FOR EMPIRICAL RANKING

The biological plausibility of this general approach rests on the significance of accumulated experience in generating useful percepts. Importantly, the accumulation that determines the empirical rank of a stimulus feature will also order perceptions of that quality. Thus in the example of line length above, every possible projected length on the retina is related to a correspondingly ordered perceptual length, albeit in a non-linear fashion. Seeing visual qualities according to their empirical rank, whether of geometrical characteristics or other qualities such as brightness or color, represents a scheme of vision that maintains in perceptual space the *relative* similarities and differences among physical objects. The result is a perceptual sense of each quality having a "proper place" in relation to the physical sources of all other such stimuli. Indeed, the strategy works so well in contending with the inverse optics problem that it is difficult to convince people that what they see on a moment-by-moment basis is not a veridical representation of what is "really out there".

SUMMARY

A high quality database of range images in which the direction and distance of every point in the images is acquired by means of laser range scanning is an essential first step in exploring geometrical perception in probabilistic terms. This information provides a straightforward way of relating the spatial attributes of the physical world and the retinal projections it gives rise to, thus providing a proxy for human visual experience. A statistical analysis of these image-source relationships can then test the hypothesis that the perceived geometry of visual stimuli (e.g., line lengths, angles, sizes, distances) is indeed based on the statistical linkage between the geometrical arrangements of elements in projected images and the arrangement of their real-world sources that humans have always witnessed.

Chapter 3

Line Length

A simple starting point in considering in greater detail whether the geometries we see are determined in a wholly probabilistic way is the apparent length of a line, or, to put the matter more generally, the perception of the spatial interval between any two points in a scene (Howe and Purves, 2002). In the absence of other contextual information, it seems logical to suppose that the percepts arising from a line of a given length (e.g., a line drawn on a piece of paper or on a computer screen) would correspond more or less directly to the proportional length in the retinal projection. Accordingly, if a series of such stimuli having different lengths were shown to observers, one would expect the apparent lengths to scale proportionally with the lengths of the retinal stimuli. This expectation, however, is not met.

VARIATION IN APPARENT LENGTH AS A FUNCTION OF ORIENTATION

An example of this discrepancy is the variation of the perceived length of the same line as a function of its orientation in the stimulus (Figure 3.1A). As investigators have repeatedly shown over the last 150 years, a line that is oriented more or less vertically in the retinal image appears to be somewhat longer than a horizontal line of the same length, the maximum length being seen, oddly enough, when the stimulus is oriented about 30° from vertical (Wundt, 1862; Shipley et al., 1949; Pollock and Chapanis, 1952; Cormack and Cormack, 1974; Craven, 1993) (Figure 3.1B).

This effect is evidently a particular manifestation of a general tendency to perceive the extent of any spatial interval differently as a function of its orientation in the retinal image. For instance, the apparent distance between a pair of dots varies systematically with the orientation of an imaginary line between them (as Wilhelm Wundt first showed in 1862), and a perfect square or circle appears to be slightly elongated along its vertical axis (Sleight and Austin, 1952; McManus, 1978). Despite extensive study of these phenomena, no generally accepted explanation has been forthcoming.

In terms of the general hypothesis about the genesis of geometrical illusions presented in Chapter 1, this variation in the perceived length of lines should reflect the

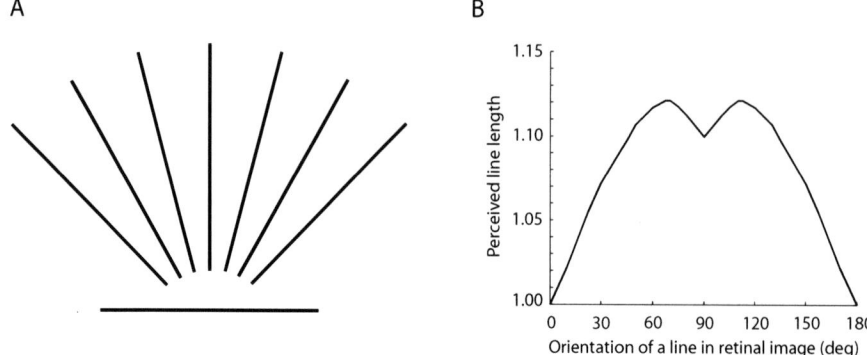

Figure 3.1 Variation in apparent line length as a function of orientation. A) The horizontal line in this figure looks shorter than the vertical or oblique lines, despite the fact that all the lines are identical in length. Notice that the horizontal/vertical comparison here is similar to the effect elicited by the T-illusion in Figure 1.2A. B) Quantitative assessment of the apparent length of a line reported by subjects as a function of its orientation in the retinal image (orientation is expressed as the angle between the line and the horizontal axis). The maximum length seen by observers occurs when the line is oriented approximately 30° from vertical, at which point it appears about 10–15% longer than the minimum length seen when the orientation of the stimulus is horizontal. The data shown here is an average of psychophysical results reported in the literature (see Pollock and Chapanis, 1952; Cormack and Cormack, 1974; Craven, 1993). (B is after Howe and Purves, 2002)

statistical relationship between the projected images of lines and the their real-world sources. Recall that the underlying rationale for this way of understanding perception is a biologically effective way of contending with the inherent ambiguity of visual stimuli. A quick review of Figure 1.1B will serve as a reminder that the projected length of a line on the retina can correspond to a line of virtually any physical length and orientation in depth in the real world.

According to the probabilistic framework outlined in Chapter 2, the apparent lengths elicited by lines projected at different orientations on the retina should be predicted by the empirical rank of the projected length of a given line on the scale determined by past human experience with the frequency of occurrence of the sources of projected lines in that orientation. Recall that the rank is specifically determined by the relative frequency of occurrence of sources that in past experience generated projections either shorter or longer but in the same orientation as the projected line in question (see Figure 2.4).

SAMPLING THE RANGE IMAGE DATABASE

To test the merits of this explanation, we sampled the physical sources of straight-line projections in the range image database described in Chapter 2. When considering lines, it is natural to think of stimuli generated by the luminance contrasts in a scene. Luminance contrast boundaries (i.e., edges) have traditionally been the focus of vision research, primarily because edges typically correspond to object boundaries, and are therefore considered "information-rich". Furthermore, there is much evidence that observers preferentially attend to contrast boundaries in normal viewing (see, for example, Yarbus, 1957). Yet another reason for the focus on edges in past research is that they generally elicit much stronger responses in visual cortical neurons than do stimuli that lack contrast boundaries.

Despite the obvious importance of edges in vision, from a purely geometrical point of view lines—straight or otherwise—are not limited to those that happen to coincide with luminance contrast boundaries. For both geometrical and behavioral purposes, a straight line is simply a set of points whose positions in space conform to a linear progression. Since understanding the perception of geometry and its behavioral consequences is the goal here, we consider straight lines in the real world, as well as other geometrical shapes and configurations examined in later chapters, as geometrical entities rather than as edges, and have conducted analysis of the range image database accordingly. This approach may seem counterintuitive in that a set of points that forms a geometrical line but does not correspond to a luminance edge does not *look* like a line. Such sets, however, are as relevant to the percepts elicited by linear stimuli as lines arising from contrast. The reason is that these implicit lines are just as important for successful navigation in typical environments as the subset of lines made explicit by contrast: for visual percepts to be useful, they must be able to guide appropriate behavior with respect to the physical sources of *any* geometrical entity, not just those that happen to correspond to luminance edges.

To appreciate this fact, consider the behavioral tasks involved in playing the children's game called "jacks". For those not familiar with the game, the challenge is to pick up a number of small objects off a surface and catch a rubber ball before it bounces a second time. Successfully picking up multiple jacks during the brief interval the ball is in the air demands exquisite visually guided behavior predicated on an appreciation of the spatial arrangement of the jacks scattered across the playing surface (in addition to good motor skills, of course). Children (and willing adults) have no problem understanding the relevant geometrical relationships (e.g., the lengths and orientations of the "lines" between the jacks), even though these sets of points on the relevant surface produce no luminance contrast boundaries.

Thus, to fully understand the perceptual response to any geometrical stimulus, the statistics of all the geometrical arrangements that are the same as or similar to the stimulus at issue must be taken into account, regardless of whether or not they are associated with edges. Accordingly, when sampling straight lines in the image database, we simply searched for any set of points that formed a straight line geometrically. Since we had no way of determining the relative influence of explicit versus implicit lines in accumulated experience with lines, the same status was given to all geometrically straight lines in the subsequent analysis, despite the fact that people look preferentially at edges.

ANALYZING THE SOURCES OF STRAIGHT LINES

Based on this general perspective, straight-line templates such as those shown in Figure 3.2A were applied to randomly selected regions of the range images described

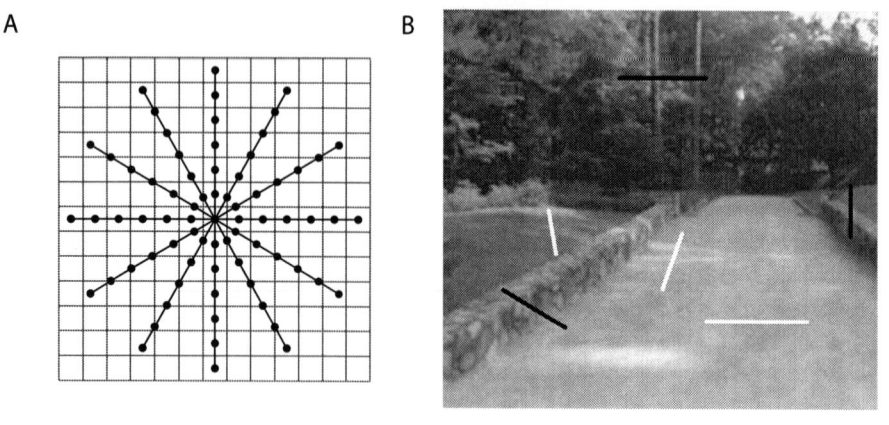

28 Perceiving Geometry: Geometrical Illusions Explained by Natural Scene Statistics

Figure 3.2 Sampling straight lines in the image database. A) The pixels in a region of one of the images in the database are represented diagrammatically by grid squares; the connected black dots indicate a series of templates for sampling straight lines at different projected orientations. The points comprising each template were evenly spaced (the distance in the image between each pair of neighboring points was one pixel); since the images in the database comprised discrete pixels, the points in the template, when overlaid on the images, did not correspond exactly to particular pixels except when the templates were horizontal or vertical. The range (distance) information associated with the points selected by the template was generated by interpolation. B) Examples of straight-line templates overlaid on a typical image. White templates indicate sets of points that corresponded to straight lines in 3-D space, and were thus accepted as valid samples of straight lines. Black lines indicate sets that failed to meet this criterion and were therefore rejected.

in Chapter 2 (see Howe and Purves, 2002 for details). The set of points underlying the template was then screened to determine if it corresponded to physical points that formed a straight line in the 3-D world. The coordinates of the points in 3-D space were determined from the range information, and a straight line fitted to these physical points using the least-squares method (Figure 3.2B). If the average deviation of these points from the fitted line was less than an arbitrary standard, the set of points was accepted as a valid sample of the source of the straight-line projection represented by the template.

To sample the frequency of occurrence of the physical sources of linear projections at different orientations in the image plane, straight-line templates of different lengths (2 to 256 pixels) were applied to the projected images in orientations that varied from 0° (horizontal) to 175° in 5° increments. Each template of a given orientation and length was systematically applied to different regions of the range images, yielding about 1.2×10^8 valid samples of the physical sources of straight-line projections. By tallying the number of these samples as a function of projected length and orientation, we determined the frequency distributions of the real-world sources of line projections in different orientations. Normalizing each of these frequency distributions then generated the corresponding probability distribution of the occurrence of the physical sources of the projected lines. Examples of the distributions generated by the analysis of the fully natural scenes in the database are shown in Figure 3.3A.

Line Length

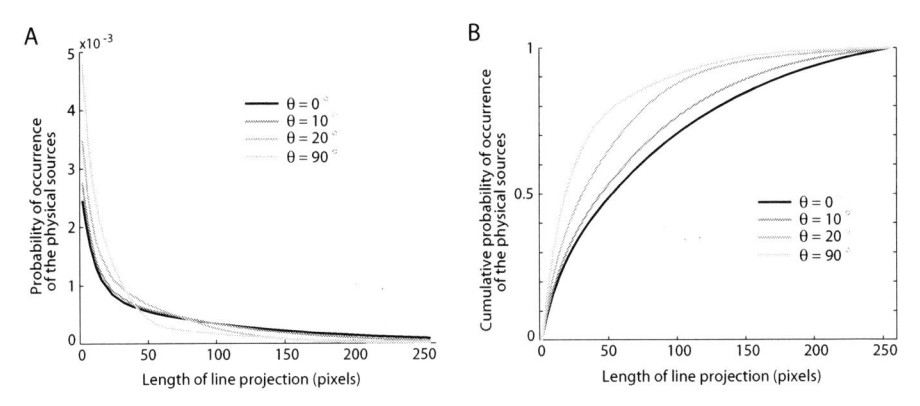

Figure 3.3 Analysis of the physical sources of straight-line projections in range images of fully natural scenes. A) Probability distributions of the occurrence of the physical sources of line stimuli. The probabilities of occurrence of the physical sources are plotted as functions of the projected length (l) of line stimuli orientated at four different orientations (θ). B) Cumulative probability distributions derived from the distributions in (A). The cumulative probability value for a given point x on the abscissa was obtained by calculating the area underneath the curve lying to the left of a line that corresponded to $l = x$ in the relevant probability distribution in (A).

EMPIRICAL RANKING OF LINE LENGTHS

Each of the probability distributions of the physical sources of lines derived in this way provides a basis for generating the empirical scale of line length pertinent to lines projected at a specific orientation. As described in the previous chapter, the purpose of this exercise is to indicate, with respect to any given line in an image, what percentage of the possible physical sources of a line projected in the same orientation generated projections shorter than the line stimulus at issue, and what percentage gave rise to longer line projections in the experience of human observers.

To determine the empirical rank of a line in the image we computed the cumulative probability distributions of the sources of lines from the probability distributions in Figure 3.3A (Figure 3.3B). A cumulative probability value is the summed probability of occurrence of all the physical sources of the linear projections that have the same orientation and are equal in length or shorter than a given projected line. Each cumulative probability distribution provides an empirical scale on which a particular projected line length in a given orientation can be ranked. It is apparent in Figure 3.3B that these empirical functions are different for lines projected at different orientations, and thus that the rank of a given line length will vary as a function of its orientation.

In terms of the empirical framework of vision outlined in Chapters 1 and 2, the perceived length of a line projected at any given orientation is determined by the percentile rank of the projected length on the relevant empirical scale. The reason is that visual percepts—in this case perceptions of length—must be ordered appropriately over the full range of possible subjective experience. Although a ruler does this perfectly well in case of physical length, the visual system cannot simply measure physical lengths because of the inverse optics problem (see Figure 1.1B); the best it can do is to order the subjective sense of length according to feedback from accumulated experience, an ordering that is useful in guiding behavior.

PREDICTION OF APPARENT LENGTH BASED ON EMPIRICAL RANK

Given this strategy of vision, the puzzling variation in the apparent length of a line as a function of its orientation in the image plane illustrated in Figure 3.1 should be predicted by the changing empirical rank of a line as its orientation changes. The relevant rank can be readily determined from cumulative distributions such as those in Figure 3.3B.

Consider, for instance, a projected line 7 pixels in length oriented at 20° (this length corresponds to ~1° of visual angle and has been used in many psychophysical studies). The cumulative distribution of the sources of lines oriented at 20° gives a cumulative probability value of 0.1494 for a line of this length. Thus 14.94% of the physical sources of lines oriented at 20° generated projections equal to or less than 7 pixels in length, and 85.06% generated longer lines. Thus, the percentile rank of a line of this projected length oriented at 20° is 14.94.

The empirical ranks of lines 7 pixels in length at different orientations ranging from 0–180° can be similarly determined from the relevant cumulative probability distributions. Figure 3.4 shows that these rankings vary systematically as a function of line orientation. Thus a line 7 pixels in length oriented vertically (90°) in the image plane holds a much higher rank on the relevant empirical scale than a line of the same length oriented horizontally. Moreover, for projected lines of the same length, the lines with the highest ranks in the pertinent empirical scales are those oriented about 20–30° from vertical. It should be obvious that the shape of the function described by these

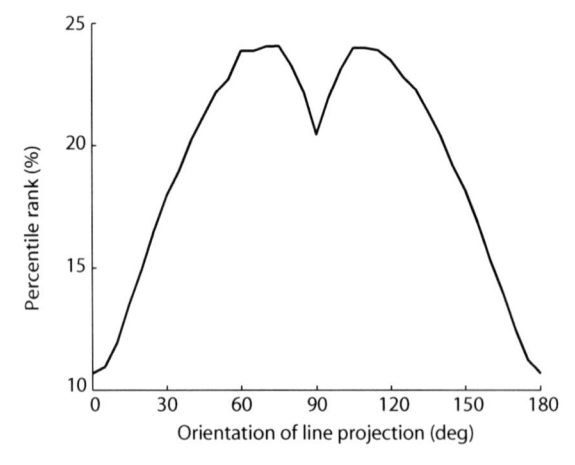

Figure 3.4 The percentile rankings determined for lines 7 pixels in length in different orientations, derived from the cumulative probabilities of the physical sources of linear projections in different orientations. This function predicts the perception of line length as a function of orientation, and should be compared with the psychophysical reports of perceived line length shown in Figure 3.1B.

Line Length

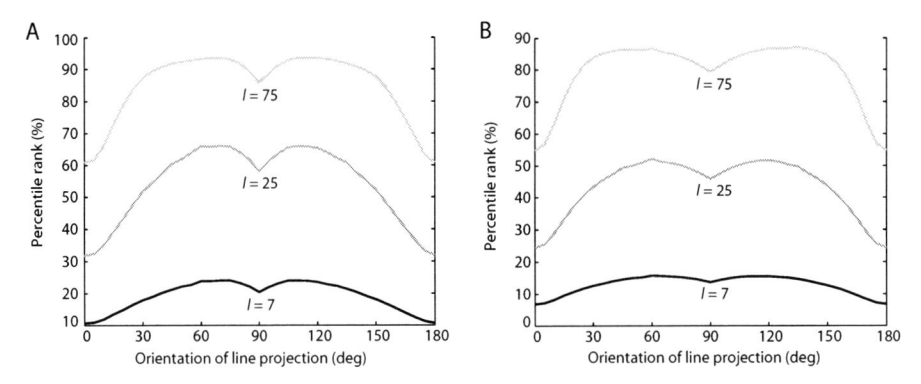

Figure 3.5 Rankings of lines of three different projected lengths (l) derived from the statistics of the fully natural scenes (A) compared to rankings derived from scenes that included human constructions (B).

empirical rankings in different orientations in Figure 3.4 is remarkably similar to the psychophysical function of perceived line length for visual stimuli of roughly this size illustrated in Figure 3.1B. Such ranking functions for projected lines of other lengths examined (Figure 3.5A) show the same general result. All these functions share the same form, with minima at 0° and 180°, maxima between 20–30° from vertical, and a dip in the function at 90°.

These results are derived from fully natural scenes in the database. The statistics from natural scenes are presumably particularly relevant to understanding human perception, since humans have, for the most part, evolved in such environments. Nevertheless, as shown in Figure 3.5B, scenes that contain human constructions ("carpentered environments") give much the same result.

Taken together, these observations show that a higher empirical rank of a line predicts a perception of longer length, thus accounting for the otherwise puzzling way in which the apparent length of a line varies as a function of its orientation. Moreover, this perceptual variation is evidently a consequence of a general statistical characteristic of the world, and does not depend on specific experience in "carpentered" environments, as has sometimes been implied.

BASIS FOR DIFFERENCES IN EMPIRICAL RANK AS A FUNCTION OF ORIENTATION

The upshot of this analysis is that the empirical rank of the length of any line varies as a function of orientation because the empirical scales of projected line length are different for differently orientated lines. What, then, is the nature of the physical bias that gives rise to these systematic differences?

To answer this question, consider the probability distributions from which the empirical scales (i.e., the cumulative distributions) are derived. In Figure 3.3A, for instance, there is a progressive change among the distributions for lines oriented at 0°, 10°, 20° and 90°, respectively: as the orientation becomes more vertical, the probability of finding physical sources for relatively short lines increases, whereas the

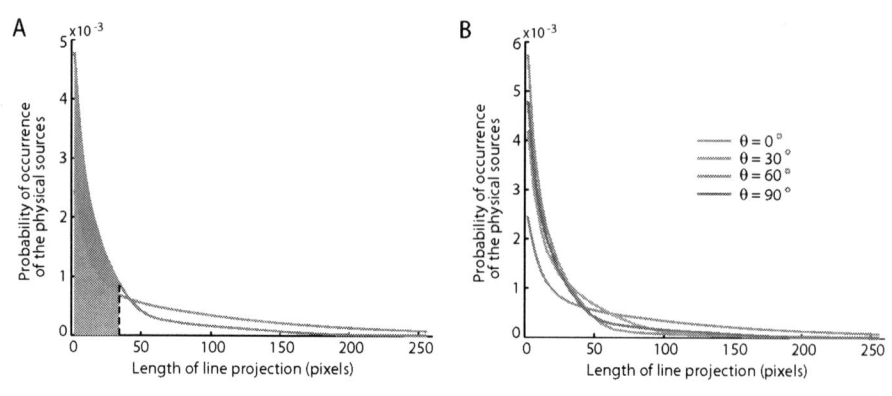

Figure 3.6 Differences in the probability distributions of the physical sources of lines as a function of orientation are the basis for the different empirical ranks of the same line length in different projected orientations (see Figure 3.5). A) Probability distributions of the sources of horizontal lines (0°, blue) and vertical lines (90°, red). The dashed line indicates a particular line length. Comparing the areas to the left of the dashed line, the area underneath the distribution for the projected line at 90° is greater than the area underneath the distribution at 0°. This difference means that the percentage of the physical sources of vertical lines that generate projections shorter than a vertical line of the length indicated is greater than the percentage of the sources of horizontal lines that generate projections shorter than a horizontal line of the same projected length. This statement holds whatever the length of the projection, although the difference to the left of the dashed line becomes smaller as line length increases. B) Probability distributions of the sources of lines at 0°, 30°, 60° and 90° (inset shows key). As line orientation is varied from 0° to 60°, the occurrence of the sources of relatively long lines decreases. This trend is reversed when the orientation varies from 60° to 90°; thus the probabilities of occurrence of relatively long lines in the distribution at 90° (red) are higher than in the distribution at 60°. The perceptual consequences are explained in the text.

likelihood of finding sources of longer lines decreases. As a result, for two stimuli that have the *same* length but different orientations (say, a horizontal and a vertical line), the percentage of all the physical sources of vertical lines that generated projections *shorter* than the vertical line in question will be larger than the percentage of sources of horizontal lines that generated shorter projections than the horizontal line in question (Figure 3.6A). In consequence, the empirical rank of the length of the vertical line among all vertical line projections is higher than the rank of the horizontal line among all horizontal line projections, even though the two lines have the same projected length. This variation of the probability distributions of the physical sources of lines as a function of orientation leads to the differences in the empirical ranks of lines that have the same projected length.

The variation of the probability distributions as the orientation of the line changes from 0° to 90° is not, however, monotonic. Figure 3.6B shows that, when line orientation is varied from 0° to 60°, the distributions change progressively, such that the occurrences of the sources of relatively long lines *decrease* as the orientation begins to move toward vertical. This trend is reversed, however, as the orientation approaches vertical; thus the sources of relatively long lines in the distribution for 90° are actually more likely than in the distribution for 60°, giving vertical lines a somewhat lower empirical rank than lines oriented 20–30° away from 90°. This statistical reversal explains why the general increase in the rank of the length of a line as a function of increased verticality peaks

Line Length

and then reverses when the line is about 20–30° from vertical, creating the maxima at these orientations and the dip at 90° evident in Figures 3.4 and 3.5.

WHY THESE BIASES EXIST

The next question is *why*, in terms of the structure of the world, the probability distributions of the physical sources of line projections vary in this systematic manner as projected orientation changes. In other words, why are there fewer sources of relatively long vertical lines compared to horizontal lines, and why, more specifically, are there even fewer sources that project as relatively long lines at 20–30° from vertical?

To understand the reasons, remember that almost all straight lines in the physical world are components of flat surfaces: a one-dimensional line, or even a close approximation thereof, rarely occurs in nature. Therefore when considering the physical sources of straight lines at different orientations, the pertinent variable is the extension of flat surfaces in different directions. Consider the physical world as a 3-dimensional Euclidean space defined by three mutually perpendicular axes, i.e., an axis that is horizontal and parallel to the retinal image plane, an axis that is vertical and parallel to the image plane, and a depth axis perpendicular to the image plane (Figure 3.7A). Horizontal line projections in the retinal image are therefore typically generated by the extension of planar surfaces in the horizontal axis, whereas vertical lines are typically generated by the extension of such surfaces in either the vertical or the depth axis.

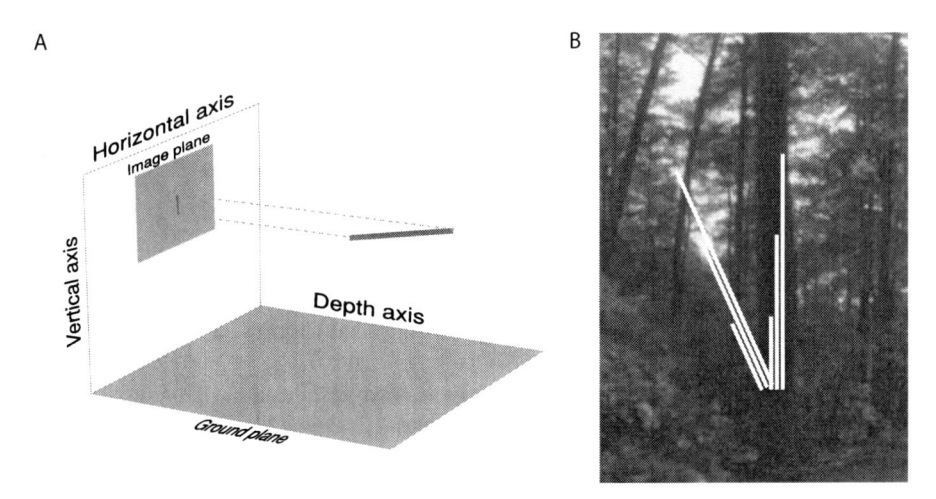

Figure 3.7 Physical basis for biases in the sources of line projections. A) Diagram of the axes that define Euclidean space. The physical source of a line (black) that extends in depth is foreshortened in projection. B) Image of a natural scene with superimposed vertical lines as well as oblique lines 25° from the vertical axis. The short, the medium and long vertical lines have, respectively, the same lengths as the corresponding oblique lines. Note that the vertical lines are generally more likely to fall on the surfaces of natural objects than the oblique lines.

A quick inspection of the world makes clear that the extension of surfaces in the vertical axis is limited by gravity. As a result, this category of the sources of vertical line projections is relatively restricted in the projected lengths they produce. The other category of the sources of vertical projections—the extension of surfaces in the depth axis—is also inherently limited in the length of projections generated because the depth axis is perpendicular to the image plane, and thus foreshortened in projection (see Figure 3.7A). Neither of these limitations presents a major restriction to the genesis of horizontal projections from real-world objects. In consequence, there are fewer sources of relatively long projected vertical lines in the world compared to sources of relatively long projected horizontal lines.

A final question concerns the real-world bias that accounts for the fact that the least number of sources are found for relatively long linear projections at 20–30° from vertical (compare the distributions at 60° and 90° in Figure 3.6B). To understand this peculiarity, consider vertical lines and lines oriented at 20–30° from vertical superimposed on images of the natural scenes (Figure 3.7B). Vertical lines are somewhat more likely to correspond to object surfaces than are lines oriented at 20–30° from vertical, simply because trees and other objects in the physical world more frequently extend along the vertical axis than they do along an axis oriented 20–30° away from vertical. As a result of the way that upright objects efficiently contend with gravity, the physical sources of relatively long linear projections at 20–30° from vertical are less probable than the sources of equally long vertical projections.

In sum, the unequal extension of surfaces in different directions is the fundamental reason why the probability distributions of the physical sources of lines are different for linear projections at different orientations, and thus why the empirical ranks of line lengths vary as a function of line orientation in the particular way they do.

PREVIOUS EXPLANATIONS OF THE ANOMALOUS PERCEPTION OF LINE LENGTH

The discrepancies between the measured lengths of line stimuli and their perception (see Figure 3.1) have, in the past, been rationalized in several different ways, including asymmetries in the anatomy of eye (Kuennapas, 1957; Pearce and Matin, 1969; Prinzmetal and Gettleman, 1993), the ergonomics of eye movements (Wundt, 1862; Luckiesh, 1922), and cognitive compensation for the foreshortening of vertical lines (Gregory, 1974; Girgus and Coren, 1975; Schiffman and Thompson, 1975; von Collani, 1985).

The last of these theories is the one most often cited as a plausible explanation. In this scenario, vertical lines in the image plane are assumed to frequently be objects on the ground that extend in depth; horizontal lines, on the other hand, are taken to be objects that are more often parallel to the frontal plane. Since lines that extend in depth are foreshortened in projection (see Figure 3.7A), the idea is that this unconscious association would cause observers to "compensate" by seeing vertical lines as longer than horizontal ones. This account, however, does not deal statistically with the fact that both vertical and horizontal lines, or lines in any orientation in the image plane, can be generated by physical sources that have any inclination in depth. The theory also provides no explanation for the psychophysical function actually observed, in particular why lines oriented at 20–30° from vertical are seen as being longer than lines at 90° (see Figure 3.1B).

Perhaps the most sophisticated approach to explaining these anomalies of apparent length is based on an analysis of the density of luminance contrast transitions in natural images (Craven, 1993). The study found that the prevalence of these so-called "zero-crossings" in images is different along different orientations, and that this variation correlates reasonably well with the variation in perceived line length as a function of orientation shown in Figure 3.1B. The author thus proposed that the visual system "calibrates" perceived length according to the density of luminance contrast transitions along a given orientation. Although the analysis of zeros-crossings is no doubt correct, there is no obvious reason why the visual system should carry out a calibration of this sort.

In fact, the physical basis for the differences among the probability distributions of the sources of lines summarized here nicely explains the correlation between the density of luminance contrast transitions and perceived line length. The frequency of occurrence of the sources of straight lines in different orientations is determined by the extension of planar surfaces in different directions. Since luminance contrast transitions in images are usually caused by interruptions in the continuity of object surfaces (by the edge of a surface or by occlusion, for example), the density of luminance contrast transitions along any given orientation is, in general, inversely related to the extension of flat surfaces in that orientation. This relationship means that for an orientation in which the average extension of real-world surfaces is relatively less—which increases the empirical rank of a given line length and therefore the apparent length of the line (see above)—the density of luminance contrast transitions would be relatively high. Thus, the reason for the correlation between perceived line length and the density of luminance contrast transitions is straightforward; by the same token, however, variation of the density of zero-crossings is not the cause of the apparent variation in line length with orientation.

SUMMARY

The observed variation in the apparent length of lines as a function of their projected orientation agrees remarkably well with the percentile ranks of lines on the relevant empirical scales of line length derived from the probability distributions of the physical sources of line projections. Thus, this otherwise puzzling peculiarity about one of the simplest aspects of perceived geometry can be neatly explained as a particular manifestation of a broader visual strategy that generates percepts according to the probability distributions of the possible sources of inherently ambiguous stimuli. These observations introduce not only the utility of the empirical ranking of a stimulus attribute in predicting percepts, but the deeper biological significance of generating percepts in this way. Both as a method of assessing the empirical significance of stimuli and as a conceptual framework for understanding vision, this theme carries on through the remainder of the book.

Chapter 4

Angles

A second fundamental aspect of the physical arrangement of objects in space and the perceptions of this geometry is the angle made between two lines that meet—either explicitly or implicitly—at a point. Like the apparent length of lines, an intuitive expectation about the perception of angle subtense is that such a basic feature of what we see should scale with the dimensions of the angles projected in retinal images. It has long been known, however, that this is not what people see. What follows is a review of the evidence that the peculiar way we see angles is, in fact, a further manifestation of the way the visual system contends with the inverse optics problem. Since the physical geometry corresponding to a given angle projected on the retina can be any of an infinite number of real-world angle subtenses and orientations in space, and since observers must deal with physical sources, the argument is again that this category of geometrical percepts is generated on an entirely statistical basis.

THE PERCEPTION OF ANGLES

A glance at Figure 4.1 makes plain that the perceived subtense of an angle can be greatly at odds with the angle presented in the stimulus; indeed, people looking at this sort of illustration find it difficult to believe that the subtenses of the angular objects in the image do not actually differ by several tens of degrees.

Even in the absence of the rich contextual information that biases the probable sources of the projected angles in this obvious way, observers tend to overestimate the magnitude of acute angles and underestimate obtuse ones by a few degrees (Figure 4.2A). This subtle yet robust phenomenon was first reported by Wilhelm Wundt (1862) and has been confirmed by a number of modern studies (Fisher, 1969; Maclean and Stacey, 1971; Carpenter and Blakemore, 1973; Heywood and Chessell, 1977; Greene, 1994; Nundy et al., 2000).

This anomalous perception of angles in a very spare presentation of angular stimuli is easiest to appreciate (and is most often demonstrated) in a closely related series of classical geometrical illusions that involve intersecting lines in various configurations. The simplest of these is the tilt illusion, in which a vertical line in the context of

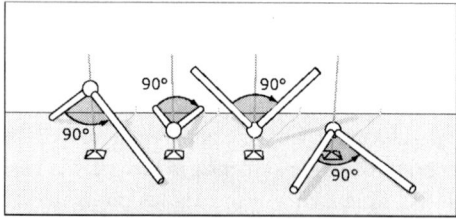

Figure 4.1 The subtenses of the angular objects in this stimulus appear to differ by as much as 60° or more. In fact, each of the objects subtends exactly the same right angle in the image (see inset). (After Purves and Lotto, 2003)

an obliquely oriented line appears to be rotated slightly away from vertical in the direction opposite the orientation of the oblique "inducing line" (Figure 4.2B). The direction of the perceived deviation of the vertical line is consistent with the perceptual enlargement of the acute angles in the stimulus, and/or a reduction of the obtuse angles. This relatively small effect is enhanced in the Zöllner illusion, which is essentially a more elaborate version of the tilt effect achieved by an iteration of the oblique stimulus elements (Figure 4.2C). The several parallel vertical lines in this presentation appear to be tilted away from each other, again in directions opposite the oblique orientation of the contextual line segments. A further well known permutation of such effects is the Hering illusion mentioned earlier (see Figure 1.2B), in which two parallel straight lines appear bowed in the context of intersecting lines whose orientations change progressively (Figure 4.2D).

The issue considered here is whether the statistical relationship between images and sources in natural scenes can also rationalize the perception of angle subtense, as well as more complex effects elicited by angular stimuli. The evidence presented below indicates that the perception of angles and all the various effects shown in Figure 4.2 are, like the perception of line length, manifestations of an empirical strategy of vision in which the angle actually seen is determined by the relative frequency of occurrence of the possible sources that human observers have found to underlie angle projections on the retina. Perceiving angles in this way allows observers to contend with the inevitable ambiguity of projected angles.

Angles

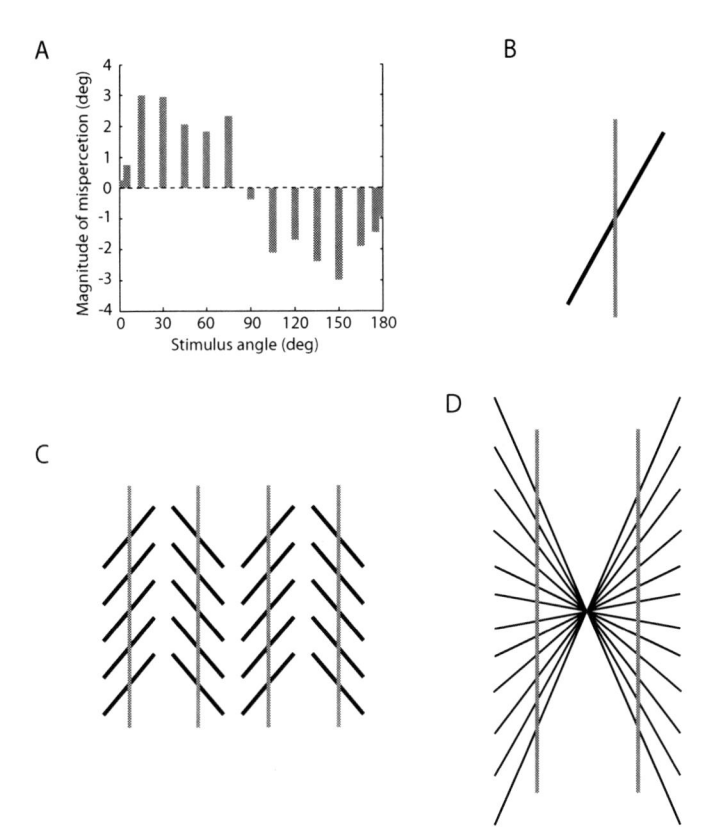

Figure 4.2 Geometrical illusions caused by stimuli in which angles between intersecting lines are a major feature. A) Psychophysical results (Nundy et al., 2000) showing the systematic misperception of angles presented simply as two lines that meet at a point on an otherwise empty field: when observers view such stimuli acute angles are slightly overestimated and obtuse ones underestimated. B) The tilt illusion. A vertically orientated test line (gray) appears to be tilted slightly counterclockwise in the context of an oblique "inducing" line rotated clockwise (black). C) The Zöllner illusion. In the standard presentation of this effect, the vertical test lines (gray) appear tilted in directions opposite to the orientations of the contextual lines (black). D) The Hering illusion. The two vertical lines (gray) appear bowed when presented in the context of radiating lines. (After Howe and Purves, 2005a)

THE PROBABILITY DISTRIBUTION OF ANGLE SOURCES

In much the same way that the physical sources of straight lines were sampled from the range images, the physical sources of angle projections can be identified using appropriate geometrical templates applied to scenes in the range image database described in Chapter 2. The first step in this process is to find regions of the scenes in the database that contain a valid physical source of one of the two lines (subsequently referred to as the *reference line*) by applying to the images a straight-line template at different orientations (Figure 4.3). If the set of points underlying the reference line template in the image corresponded to physical points that formed a straight line in 3-D space, the physical points were accepted as a valid source of the reference line.

40   Perceiving Geometry: Geometrical Illusions Explained by Natural Scene Statistics

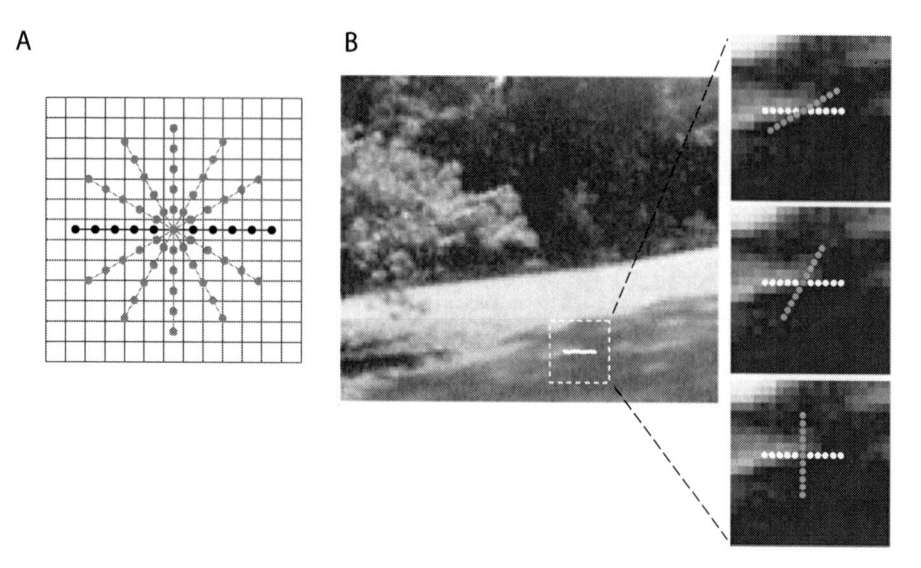

Figure 4.3 Sampling the physical sources of angles. A) The pixels in an image region are represented diagrammatically by the grid squares. The black dots indicate a reference line template and the gray dots a series of additional templates for sampling a second line oriented at various angles with respect to the reference line (only a few are shown; the actual sampling entailed all the angles between 0 and 175° in 5° increments). The reference line template was also oriented at 45° or 90° to test oblique and vertical as well as horizontal. B) The set of white points overlaid on the image indicates a valid sample for the reference line in (A). Blowups of the boxed area show examples of the second template (gray) that was overlaid on the same area of the image to sample for the presence of a second straight line in different orientations; in each of the cases shown, the second template also identifies a valid sample. (After Howe and Purves, 2005a)

When a valid physical source of the reference line was found in a region of a scene, the probability of occurrence of a second line forming an angle with the reference line in the same region was then determined. As illustrated in Figure 4.3, this further assessment was made by overlaying an additional straight-line template in a series of different orientations on the image sample, asking whether the points underlying the additional template also corresponded to a straight line in the 3-D space. By systematically applying these templates to the images, the relative frequency of occurrence of the physical sources of angles with specific projected subtenses could be tallied.

The results of sampling range images in this manner are shown in Figure 4.4. Regardless of the different orientations of the reference line (indicated by the black line in the icons under the graphs) or the type of scene from which the statistics are derived, the probability distributions of angle sources form a trough with lower probability values for the physical sources of angle projections that approach 90°. In other words, at any orientation of the reference line, the probability of finding real-world sources of an intersecting second line decreases as the two lines become increasingly orthogonal. This outcome applies both to scenes that are fully natural and those scenes in the database that contained some or mostly human artifacts (cf. upper and lower rows in Figure 4.4). These results are consistent with the finding that given a contour at a particular orientation, nearby contours are likely to be collinear with that contour (Geisler et al., 2001).

Angles

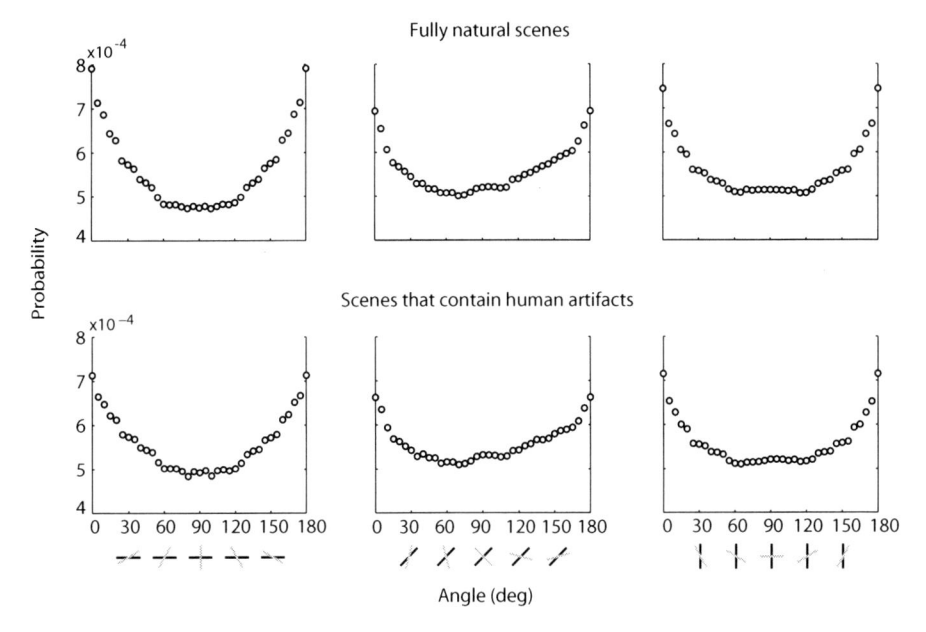

Figure 4.4 The probabilities of occurrence of the physical sources as a function of the subtense of the projected angles. The three columns represent the physical sources found using a horizontal (left), oblique (middle) or vertical (right) reference line, as indicated by the icons below the graphs. The upper row represents the results obtained from fully natural scenes and the lower row from environments that contained some or mostly human artifacts. (After Howe and Purves, 2005a)

SOURCE OF THE BIAS

The bias evident in these statistical observations can be understood by considering the provenance of straight lines in the physical world. As noted in the last chapter, almost all straight lines in the real world are components of planar surfaces. A region of a planar surface that contains two physical lines whose projections intersect at 90° would, on average, be larger than a surface that included the source of the two lines of the same length that are less orthogonal to each other (Figure 4.5). Since larger surfaces include smaller ones, the probability of finding larger planar surfaces in the world is necessarily lower than the probability of finding smaller ones. Thus, other things being equal, the occurrence of the physical sources of angles that project at or near 90° is statistically less likely than the occurrence of the sources of angles that are nearer 0° or 180°.

PREDICTING THE PERCEPTION OF ACUTE AND OBTUSE ANGLES

Since the results derived from different types of scenes, as well as the results using reference lines at different orientations are similar, the data in Figure 4.4 were pooled for further analysis. Figure 4.6A shows the cumulative probability distribution of angle sources derived from this overall distribution, which gives the summed probability

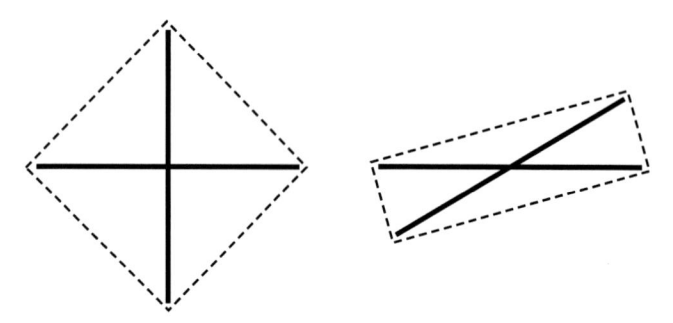

Figure 4.5 The physical source of two lines intersecting at or near 90° is likely to be part of a planar surface (dashed line) that is larger, on average, than a surface that contains the source of two less orthogonal lines.

of occurrence of the sources of all the projected angles that are less than or equal to a given subtense. Much like its significance in understanding the perception of line length, the cumulative distribution function is an empirical measure of past experience with angle subtenses: for any given subtense, the corresponding cumulative probability value indicates the percentage of the physical sources of all angles that project as angles smaller than the projected subtense in question, and the percentage that generate larger angle projections.

For example, if the angle under consideration is 30°, then the corresponding cumulative probability value on the gray curve in Figure 4.6A is 0.185; this means that ~18.5% of the physical sources of angle projections generated projected angles equal to or less than 30°, and that ~81.5% generated angles larger than 30°. When compared to a cumulative distribution derived from a hypothetical probability distribution in which the probability for the physical sources of all angles is uniform (indicated by the black line in Figure 4.6A), the cumulative distribution of angle sources derived from the image database gives somewhat higher values for angles less than 90°, and somewhat lower values for angles greater than 90°.

If, as our hypothesis about visual space supposes, perceptions of angle subtense are generated probabilistically on the basis of past human experience, then the angles seen should accord with their relative ranking on the empirical scale of angle subtenses defined by the cumulative probability distribution of the sources of angles. If the probability of the sources of all angle subtenses were uniform, the ranks of all angles would be evenly distributed on this empirical scale, i.e., an angle of x degrees would have always corresponded to a cumulative probability of x/180 (as indicated by the black line in Figure 4.6A). As a result, the rank of an angle x on the empirical scale would always be the same as the rank of the angle on the linear scale of 0° to 180°, and the perceived subtense of any angle predicted in this way would always match its actual subtense. The actual distribution of the occurrences of the physical sources of angles derived from the image database, however, is not uniform (see Figure 4.3 and the gray curve in Figure 4.6A). As is apparent in Figure 4.6A, the cumulative probability for any angle x between 0° and 90° is somewhat greater than x/180, meaning that the rank of any such angle on the empirical scale is shifted slightly in the direction of 180°

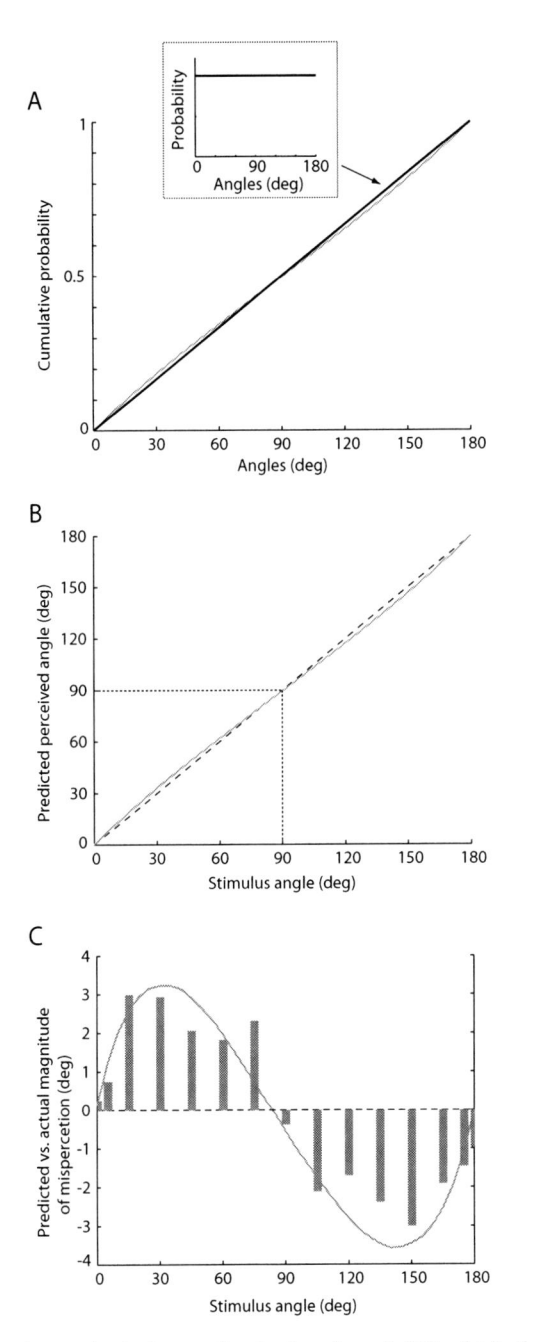

Figure 4.6 Predicting the perceived subtense of angles from the probability distributions of angle sources in the database. A) The gray curve is the cumulative probability distribution of the physical sources of angles derived from the probability distribution pooled from the 6 distributions shown in Figure 4.4. The black line, in contrast, indicates the cumulative probability distribution derived from a hypothetical distribution in which the probability of any given source is the same (see inset). B) The predicted perceptions of angle subtenses follow from the empirical ranks of angles in the cumulative probability distribution and are indicated by the gray curve; the dashed black line indicates the actual subtenses of the stimuli. C) The magnitude of angle misperception predicted by the analysis (i.e., the difference between the gray and the black lines in [B]; indicated by the gray curve) compared to psychophysical measurements of angle misperception shown in Figure 4.2A. (After Howe and Purves, 2005a)

compared to its position in the geometrical space of 0 to 180°. The opposite is true for any angle between 90° and 180°.

To illustrate this point more specifically, consider again an angle of 30°. The position of this angle in the geometrical space of 0 to 180° is 30/180. The cumulative probability corresponding to 30°, however, is 0.185 ($\approx$33/180), which is larger than 30/180. This means that a 30° angle ranks higher than 30/180 among all angle projections experienced by human observers. In contrast, any obtuse angle of x degrees would rank lower than x/180 on the empirical scale of angle subtenses. Thus the ranks of both acute and obtuse angles on the empirical scale of angles are shifted systematically toward 90° compared to their positions in the geometrical space of 0 to 180°.

As indicated in Figure 4.6B, the predicted subtenses of the perceived angles are given by the cumulative probability for any angle x multiplied by 180. Accordingly, the subtense of any angle between 0° and 90° should be systematically enlarged in perception, whereas the subtense of angles between 90° and 180° should be reduced. A comparison of the angle misperceptions actually seen by subjects and those predicted by this analysis shows remarkably close agreement (Figure 4.6C).

EXPLANATION OF THE TILT, ZÖLLNER AND HERING ILLUSIONS

The several other classical geometrical illusions illustrated in Figure 4.2 can also be explained in this framework. Consider, for example, the apparent tilt of a vertical line caused by the presence of an oblique line that intersects it (see Figure 4.2A). In this instance, the perceptual effect is predicted by the probability of occurrence of the physical sources of a second line oriented at various angles, given a reference line oriented at 60° from the horizontal (Figure 4.7A). The orientation of this specific reference line was chosen because an inducing line at 60° is frequently used to demonstrate the tilt effect; the argument that follows, however, applies to an "inducing" line at any orientation. The position of a vertical line (i.e., a line rotated 30° from the reference line) is indicated by the dashed line in this distribution. The cumulative probability value associated with the 30° angle in such a stimulus is 0.184, which, when multiplied by 180, predicts that the perceived angle between the reference line and the vertical line should be 33.2°, or 3.2° greater than the actual angle between the two lines (the same argument can be made if the obtuse rather than the acute angle between the two lines is considered). Accordingly, the vertical line in the context of a line oriented at 60° should be perceived as being rotated away from the reference line slightly more than it actually is, thus appearing not quite vertical (Figure 4.7B). This prediction is again consistent with what observers see in response to this sort of stimulus (see Figure 4.2A and Bouma and Andriessen, 1970; Wenderoth et al., 1979; Greene, 1994).

The Zöllner illusion (see Figure 4.2C) is essentially an iteration of simple tilt stimuli, producing an overall effect in which the vertical test lines appear to be more markedly tilted away from the contextual lines. Similarly, the parallel lines in the Hering stimulus (Figure 4.2D) appear bowed because of the concatenation of tilt stimuli. In this case, the upper and lower contextual lines tilt in opposite directions, causing the perceived orientation of corresponding components of the test lines to change progressively, resulting in the apparent bowing.

Angles

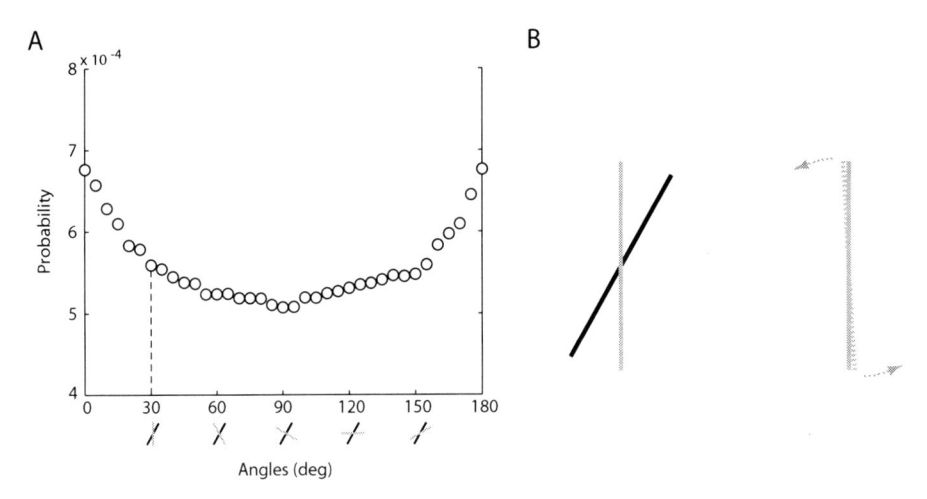

Figure 4.7 Statistical explanation of the tilt illusion. A) Probability distribution of the physical sources of projected angles in the database, given a reference line oriented at 60° from the horizontal (indicated in black in the icons below). B) Left panel shows the standard presentation of the tilt illusion; the gray line is vertical. Right panel indicates the direction (arrows) and magnitude of the tilt effect predicted by the distribution in (A). The solid gray line is vertical; the dotted line indicates the predicted shift in the apparent position of the vertical line in the left panel. (After Howe and Purves, 2005a)

NEURAL MECHANISMS UNDERLYING THE PERCEPTION OF LINES AND ANGLES

In addition to rationalizing the otherwise puzzling effects in perceived geometry, this explanation of the perception of angles and line orientations has the advantage of providing a plausible framework for better understanding the neural mechanisms underlying these aspects of vision.

The visual circuitry relevant to processing information from simple line and angle stimuli has been studied in considerable detail. A basic finding in this work is that many neurons in visual cortex selectively respond to lines or edges at particular orientations (Hubel and Wiesel, 1959; see Figure 9.1). Moreover, in many mammals neurons with these selective properties tend to be spatially organized such that there is a systematic progression in the "preferred" orientation of the cells within any given region of the visual cortex (Hubel and Wiesel, 1968).

Based on the well-documented existence of lateral inhibition among sensory elements in the input stages of the visual system (Kuffler, 1953; Hartline, 1969), a natural suggestion has been that similar lateral inhibitory effects among these orientation-selective cells in the visual cortex might underlie the anomalous percepts illustrated in Figure 4.2 (e.g., Andrews, 1967; Blakemore et al., 1970; Bouma and Andriessen, 1970; Carpenter and Blakemore, 1973; Hotopf and Robertson, 1975). In this conception, the cortical response to an angle differs from a simple summation of the pattern of activity elicited by each angle arm alone because the orientation domains co-activated by the two arms presumably inhibit each other. The effect would be to shift the distribution of the resulting cortical activity towards orientation domains whose selectivity is more orthogonal than would otherwise be the case.

Although this idea seems to be able to rationalize the perceptual enlargement of acute angles, it does not provide a physiological basis for the underestimation of obtuse angles. A further difficulty is that, although the initial physiological work on this issue indeed found inhibitory lateral interactions among orientation-selective neurons in the visual cortex (Blakemore and Tobin, 1972; Nelson and Frost, 1978), more recent studies have shown that the spectrum of interactions among visual cortical neurons is far more complex than the simple inhibitory effects originally envisioned (see, for example, Gilbert and Wiesel, 1990; Li and Li, 1994; Dragoi and Sur, 2000). As it turns out, the effect of a contextual line on the response to a target line can be inhibition, facilitation, or some combination thereof, depending on a host of factors (e.g., the orientation and the length of the stimuli, and the other response properties of the neurons involved). As a result, there is no consensus about the interpretation of such interactions, or how they are related to the perceptions of angles and line orientations.

Given the accurate perceptual predictions based on the statistical relationship of angles and their physical sources described above, it seems more likely that these diverse cortical interactions are manifestations of the empirical associations between the projections of intersecting lines and their sources, the full spectrum of contextual interactions in visual cortex reflecting the full range of possible image-source relationships. The pattern of cortical activity in response to two intersecting lines would, in this conception, reflect the empirical distribution of the real-world sources of the intersecting lines in the retinal image. Since on the empirical scale of angle subtenses defined by this distribution the angle between two intersecting lines would always be shifted towards 90° compared to the actual angle in the stimulus (see Figure 4.6A), the peaks of neuronal activity elicited would be shifted toward domains of orientation selectivity more orthogonal than those elicited by each line alone. In addition to providing a unified framework for understanding neuronal responses to both acute and obtuse angles, this prediction offers a new way of considering the relevant neurophysiology. Instead of being an epiphenomenon of cortical processing, the altered peaks of cortical activity in this interpretation reflect the accumulated statistical information conveyed by the past experience of human observers.

SUMMARY

Although the perceptual distortions that occur when viewing acute or obtuse angles in the absence of other contextual information may seem trivial with respect to the success or failure of human behavior, the visual strategy they signify lies at the core of vision. The advantage of the probabilistic processing reflected in these perceptual anomalies is that the similarities and differences among objects in the physical world are preserved in perceptual space, ensuring that what an observer sees provides a beneficial guide to action in the face of the inevitably uncertain meaning of retinal images. Since the circuitry underlying orientation is one of the most thoroughly studied aspects of the brain, this evidence about the probabilistic nature of angle perception also suggests a way of beginning to understand how the visual system instantiates these statistics.

Chapter 5

Size

Another fundamental aspect of the perception of geometry is object size. Although the apparent length of lines and the subtense of angles are certainly pertinent to size, here we consider the appearance of object size more generally. Historically, studies of this aspect of vision have focused on yet another broad category of classical geometrical illusions characterized by effects referred to as size contrast and size assimilation.

THE PERCEPTUAL EFFECTS ELICITED BY SIZE CONTRAST AND ASSIMILATION STIMULI

These effects concern differences in the apparent size of two identical targets when they are embedded in different contexts. Several standard presentations of size contrast and assimilation stimuli are illustrated in Figure 5.1. These stimuli typically entail two identical forms surrounded by one or more larger or smaller forms, generally of the same type (e.g., a circle surrounded by circles, or a square surrounded by squares).

The most thoroughly studied size contrast stimulus is the so-called Ebbinghaus figure (Figure 5.1A), in which two identical target circles are surrounded by several larger or smaller circles, respectively. The effect of these juxtapositions is that the target surrounded by the larger circles appears a little smaller than the identical target surrounded by the smaller ones (e.g., Zigler, 1960; Massaro and Anderson, 1971; Pressey, 1977). Another aspect of the anomalous perception elicited by the Ebbinghaus stimulus is that when the diameter of the surrounding circles is kept constant, the apparent size of the central target circle decreases as the interval between the central and the surrounding circles increases (Massaro and Anderson, 1971; Girgus et al., 1972; Jaeger and Grasso, 1993) (Figure 5.1B). The same size contrast effect is elicited by a concentric arrangement of circles, in which case the stimulus is referred to as the Delboeuf figure (Luckiesh, 1922; Girgus et al., 1972) (Figure 5.1C).

The Delboeuf figure, however, has a variety of other presentations, some of which elicit a size contrast effect, whereas others give rise to a so-called size assimilation effect. Thus when concentric circles are compared to a single circle identical in size to

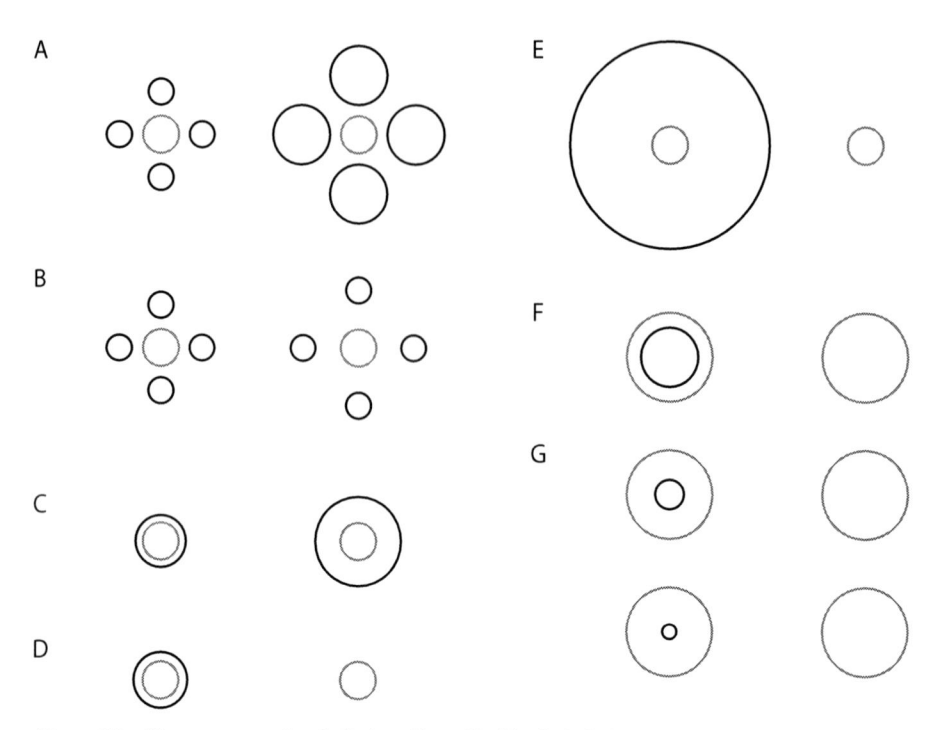

Figure 5.1 Size contrast and assimilation effects (the identical circles to be compared in each stimulus are indicated in gray). A) Standard presentation of the Ebbinghaus illusion. Observers see the central circle surrounded by smaller circles as being appreciably larger than the identical circle surrounded by larger circles. B) Even when the size of the surrounding circles in a stimulus is the same, the central circle looks smaller when the interval between the central and the surrounding circles is increased. C) A similar size contrast effect is generated by a concentric presentation, known as the Delboeuf illusion. D) The inner circle of a Delboeuf figure appears larger than a single identical circle if the diameter of the outer circle is not more than about twice that of the inner circle. E) When, however, the diameter of the outer circle is much larger than the inner circle, the inner circle looks smaller than an identical single circle. F) The outer circle of a Delboeuf figure appears smaller than a single circle of the same size. G) The effect in (F) diminishes as the inner circle becomes progressively smaller relative to the outer circle. (After Howe and Purves, 2004)

the inner target circle of the concentric set, the inner circle appears a little larger than the single circle (Obonai, 1954; Oyama, 1960; Howard et al., 1973; Pressey, 1977) (Figure 5.1D). This effect is referred to as size assimilation because the perceived size of the inner circle appears to be "assimilated" into the size of the surround (Obonai, 1954; Rock, 1995). This effect is diminished, however, when the diameter of the surrounding circle is increased. Indeed, when the diameter of the surrounding circle is sufficiently large, the overestimation of the size of the inner circle changes to a slight underestimation (Obonai, 1954; Oyama, 1960; Pressey, 1977; Jaeger and Lorden, 1980) (Figure 5.1E). Equally puzzling is the observation that the outer circle of the concentric set appears smaller when compared to a single circle of the same size (Figure 5.1F), an effect that decreases and eventually disappears as the difference between the sizes of the outer and the inner circles increases (Ikeda and Obonai, 1955; Oyama, 1960) (Figure 5.1G).

Although various explanations of the phenomena illustrated in Figure 5.1 have been proposed, there has been no agreement about the basis of size contrast and assimilation effects (Robinson, 1998). As with the phenomenology of line and angle perception, the variety and complexity of these effects has been resistant to any coherent explanation.

DETERMINING THE PHYSICAL SOURCES OF SIZE CONTRAST STIMULI

If, as implied by the evidence in the two preceding chapters, the anomalous perception of size relationships elicited by contrast and assimilation stimuli is also a manifestation of a fundamentally probabilistic strategy of vision, then the visual effects generated by the all the various configurations in Figure 5.1 should be predicted by the accumulated experience with their possible real-world sources. In this conception, the identical targets appear different in size because the probability distributions of the possible sources of the targets, given their different contexts, are different. According to the general hypothesis being examined here, it is the statistical structure of this accumulated experience that determines the geometrical characteristics ultimately seen.

To examine the merits of this supposition, we sampled the range image database to identify the physical sources that could give rise to projections whose geometrical structure was the same or similar to the size contrast or assimilation stimulus of interest (Figure 5.2). By computing the frequencies of occurrence of the physical sources of target circles embedded in each of the different contexts as a function of the projected size of the target circles, we could in this way generate the probability distribution of the sources of the targets in the context of interest. Each of these distributions thus provides the basis for constructing an empirical scale that ranks the size of a target circle in a particular context. As in the case of lines and angles, the rank of a target in a given context indicates the percentage of the physical sources of target circles that generated projections smaller than the size of the given target, and the percentage that generated larger targets. If the probabilistic framework outlined earlier is correct, then the different rankings of the size of a target on these empirical scales of target size should predict the different apparent sizes of identical targets in different contexts.

EFFECT OF CHANGING THE DIAMETER OF THE SURROUNDING CIRCLES IN THE EBBINGHAUS STIMULUS

Using this approach, we could ask how the probability distributions of the physical sources of the central target circles in the Ebbinghaus configuration vary as a function of the diameter of the surrounding (contextual) circles (Figure 5.3A).

As is apparent in Figure 5.3B, the probability of occurrence of the physical sources of the target circle decreases as the projected size of the target increases. However, this overall decline as a function of target size is obviously more pronounced when the surrounding circles are relatively small than when they are larger (compare the slopes of five curves in Figure 5.3B). In other words, the probability distributions of the

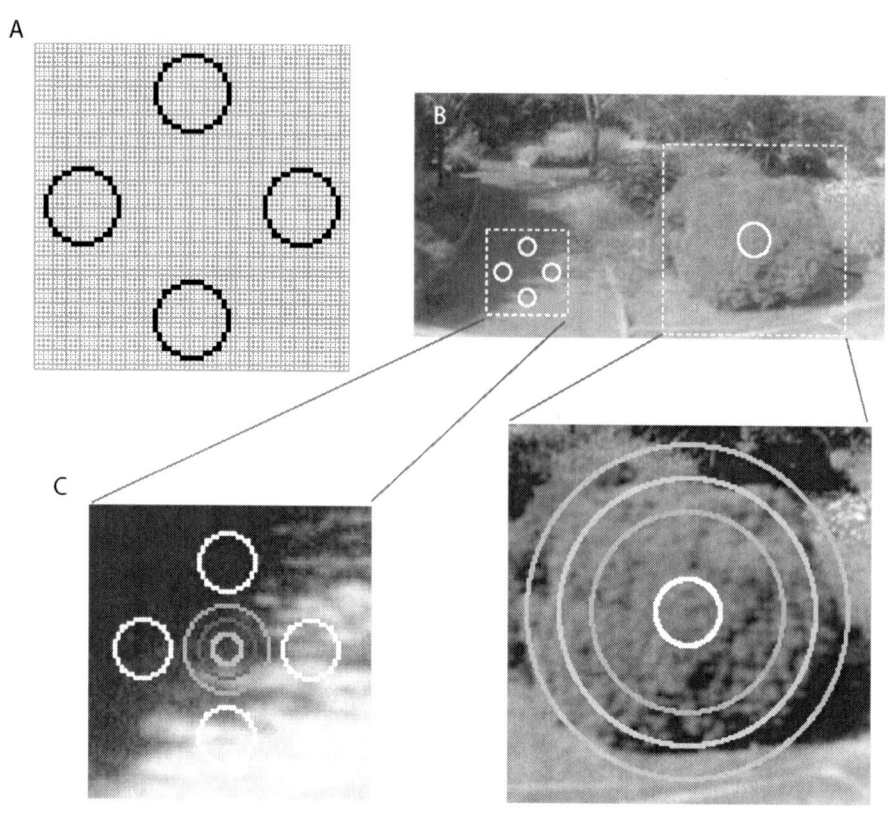

Figure 5.2 Sampling range images using size contrast and assimilation templates. A) As in earlier chapters, pixels in an image region are represented by grid squares; the black pixels indicate a template of the surrounding circles in an Ebbinghaus stimulus overlaid on the image. The pixels covered by the template thus comprise a potential sample of the contextual elements of the Ebbinghaus stimulus. If the set of physical points corresponding to the pixels comprising each of the circles formed a geometrical plane in 3-D space, the set was accepted as a valid sample of the physical source of the contextual circles. B) The pixels underlying the four template circles on the left are an example of a valid sample of this sort; pixels underlying the template on the right are a valid sample of the contextual circle (the inner circle in this case) in a Delboeuf stimulus. C) Blowups of the areas delineated by the boxes in (B). After identifying physical sources capable of giving rise to projections that appropriately matched the *contextual component* of the stimulus of interest (the white circles), we determined the frequencies of occurrence of the sources of all possible *target circles*, given the presence of the contextual components. To this end, a series of target templates of various sizes (colored circles) was sequentially overlaid on the same image region (only 3 such templates are shown here as examples). The set of pixels underlying each target template was then examined according to the same geometrical criterion applied to the contextual circles; the physical points corresponding to these pixels were counted as a physical source of the target circle only if this criterion was met. (After Howe and Purves, 2004)

sources of target circles vary systematically as a function of the size of the surrounding circles.

Given these different distributions, why then should two targets identical in size be perceived differently when surrounded by contextual circles of different sizes, as in the standard Ebbinghaus stimulus? To answer this question, consider a target circle

Size

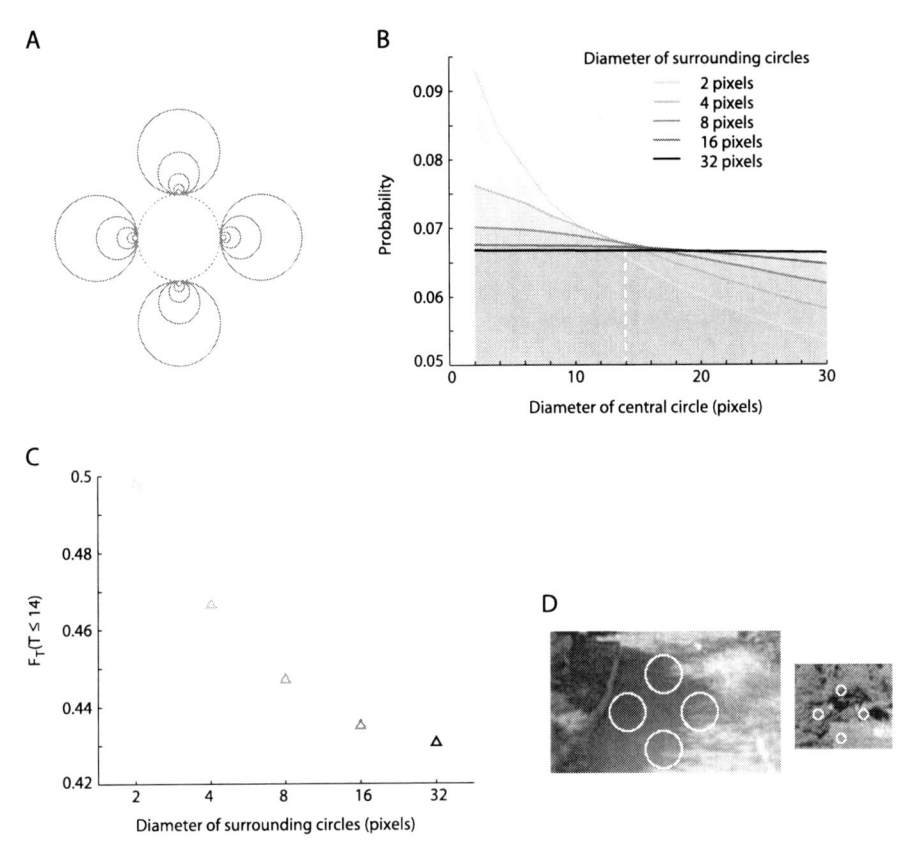

Figure 5.3 Statistical analysis of the standard Ebbinghaus stimulus. A) The relative positions of the innermost point on each of the four surrounding circles were fixed as the diameter of the circles was systematically varied; thus the largest possible size of the central target circle (the dotted circle) was the same when sampling these configurations with surrounding circles of different sizes. B) Probability distributions of the occurrence of the physical sources of target circles associated, respectively, with surrounding (contextual) circles having five different projected sizes. The probability of the physical sources of the target circle is plotted as a function of the diameter of the target in the 2-D images. The dashed line indicates, as an example, the position where the diameter of the target circle equals 14 pixels. C) The cumulative probabilities for this target size (F_T ($T \leq 14$)) derived from the distributions in (B). D) Examples of the sorts of regions in natural scenes from which a sample of large contextual circles (left) or small contextual circles (right) was likely to be obtained. (After Howe and Purves, 2004)

14 pixels in diameter as an example. The dashed line in Figure 5.3B indicates the location of this particular target size in the several different probability distributions of the sources of target circles associated with different surrounding circles. Since the ratio of the area on the left of the dashed line to the area on the right is greater for the probability distributions derived in the presence of the smaller contextual circles than the larger ones, a target circle 14 pixels in diameter occupies a different relative position in each of these probability distributions.

As in the preceding chapters, the parameter that best describes these different relative positions is the cumulative probability derived from each distribution. If T stands for target size and x the diameter of a circle in pixels, then F_T ($T \leq x$) denotes the summed or cumulative probability of occurrence of the physical sources of target circles smaller than or equal to a specific target diameter x. A greater value of F_T ($T \leq x$) means that a larger percentage of the physical sources of a projected target circle in a given set of contextual circles generated a target smaller than x pixels in diameter. Thus, the greater the value of F_T ($T \leq x$), the higher the rank of a given target size x on the empirical scale of target sizes associated with the given set of contextual circles, and the larger the target should appear.

Continuing with the example of a target circle 14 pixels in diameter, the value of F_T ($T \leq 14$) can be calculated from each of the five different probability distributions in Figure 5.3B. As shown in Figure 5.3C, the cumulative probability decreases progressively as a function of the size of the surrounding circles associated with each probability distribution. This relationship means that when the surrounding circles are small, a target 14 pixels in diameter ranks relatively high on the empirical scale of target size; conversely, when the surrounding circles are large, the same target size ranks relatively low on the pertinent empirical scale. The same logic applies to any specific target size. Thus, if the apparent size of the identical targets in the Ebbinghaus stimulus is determined by these empirical ranks, a given target circle presented in the context of small circles should appear larger than the same target in the context of large surrounding circles. This is, of course, what observers see in response to this stimulus (see Figure 5.1A).

THE PHYSICAL BASIS FOR THESE STATISTICAL DIFFERENCES

As in the case of lines or angles, an obvious question is what underlying aspects of the real world give rise to the different probability distributions of the physical sources of the target circles shown in Figure 5.3.

A circle in the retinal image can, of course, be generated by projections of an infinite variety of circles or ellipses in 3-D space. Whatever the source, be it a circle or an ellipse, the set of physical points comprising the source defines a planar surface. Thus, just as lines and angles arise from planar surface patches, the elements of the Ebbinghaus stimulus will typically have arisen from planes in the 3-D world. It follows that the probability of encountering the physical source of a target circle in the database decreases as the size of the target circle increases (see Figure 5.3B), simply because, as mentioned already, the larger planes will always encompass smaller ones.

A further question is why the presence of contextual circles of various sizes modulates the occurrence of the sources of target circles differently. Since larger surrounding circles necessarily arise from larger planes in the physical world, the relevant region of the scene is less likely to contain depth discontinuities. That is, the region will tend to comprise surface components that make up a "smoother" whole, as can be appreciated in the example shown in Figure 5.3D. A region with a smoother physical structure is more likely to contain a larger planar area capable of giving rise to the projection of a larger central target circle. In short, the presence of larger contextual

Size

circles increases the probability of occurrence of the physical sources of larger target circles. As a result, the probability distribution of the sources of target circles varies according to the size of contextual circles, as indicated in Figures 5.3B and C.

VARYING THE INTERVAL BETWEEN TARGET AND SURROUNDING CIRCLES

A further anomalous percept elicited by Ebbinghaus stimuli is that when the size of the surrounding circles is kept constant, the apparent size of the central target circle decreases as the interval between the central and surrounding circles increases (see Figure 5.1B).

Figure 5.4A shows the probability distributions of the sources of target circles in the presence of contextual circles of a constant size at various distances from the center of the stimulus. The range of possible target sizes differs among these distributions because the largest possible target size increases as the contextual circles are further separated. This difference in the maximum target size naturally causes the same target to have different relative locations in the three probability distributions, as indicated by the dashed line (an example in which the diameter of the central circle is again 14 pixels).

When the cumulative probability for a target circle 14 pixels in diameter is determined from each of these distributions, it is apparent that the empirical rank of the target decreases as the interval between the target and the surrounding circles increases (Figure 5.4B). This relationship means that when the interval is small, a target size of 14 pixels (or any value within the range of possible target sizes) will rank relatively

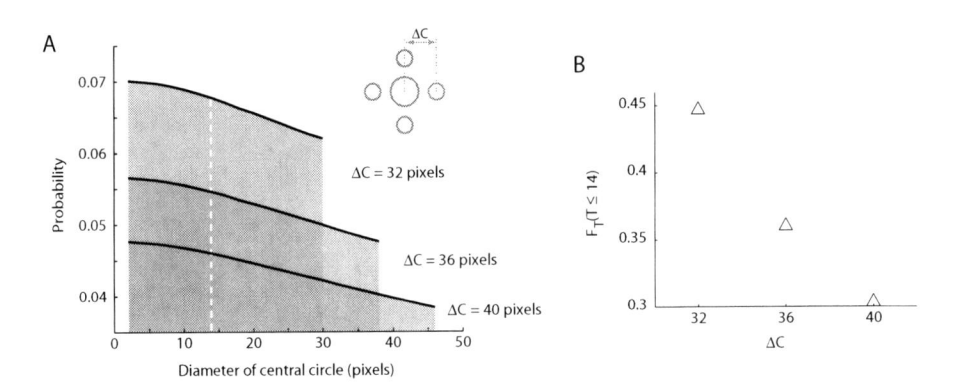

Figure 5.4 Statistical analysis of Ebbinghaus stimuli with different intervals between the central and surrounding circles. A) The probabilities of occurrence of the physical sources of target circles as functions of projected target size, given the presence of the same surrounding circles at different intervals from the center of the target (ΔC is the distance between the centers of the central and surrounding circles in the image plane; the surrounding circles are 8 pixels in diameter). The dashed line indicates, as an example, a target circle 14 pixels in diameter. B) The cumulative probabilities F_T ($T \leq 14$) derived from the distributions in (A). (After Howe and Purves, 2004)

high on the empirical scale of target size associated with that interval; conversely, when the interval is large, the same target will rank relatively low on the relevant empirical scale. This statistical relationship correctly predicts that the apparent size of the same target circle decreases as the interval between the target and the surrounding circles in the Ebbinghaus stimulus increases.

OTHER SIZE CONTRAST STIMULI

Another classical but different size contrast stimulus is the Delboeuf figure (see Figure 5.1C). When the two sets of concentric circles are compared, the inner target circle within the relatively small contextual circle appears larger than the identical target in the relatively large outer circle.

This effect can be rationalized in the same framework used to explain the perceptual responses elicited by the standard Ebbinghaus stimulus and its variants. The probabilities of occurrence of the physical sources of an inner target circle, given the presence of an outer concentric circle, are shown in Figure 5.5A. Since the largest possible target size is limited by the size of the outer circle, the location of any specific target size in these probability distributions varies according the size of the outer circle. The cumulative probabilities again show that a target circle of a given diameter ranks higher on the empirical scale associated with a smaller outer circle than on the scale associated with a larger one (Figure 5.5B). This statistical relationship thus predicts that the target within the smaller outer circle should appear larger than the same target in the larger outer circle, as is the case.

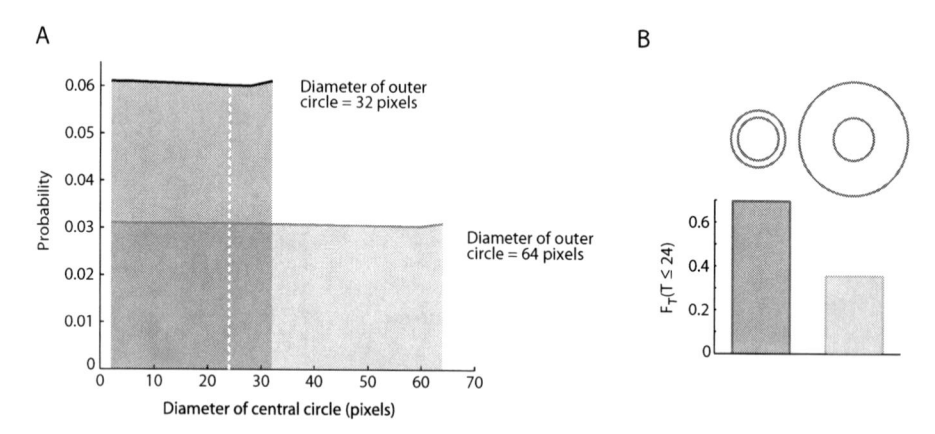

Figure 5.5 Statistical analysis of a Delboeuf stimulus, which comprises two sets of concentric circles. A) The probability of occurrence of the physical sources of inner (target) circles in the presence of an outer (contextual) circle as a function of the diameter of the target circle in the image plane. Probability distributions associated with contextual circles of two different sizes are shown. The dashed line indicates, as an example, the position in the two distributions of a target circle 24 pixels in diameter. B) Cumulative probabilities for a target circle 24 pixels in diameter (i.e., F_T ($T \leq 24$)) derived from the two probability distributions in (A). (After Howe and Purves, 2004)

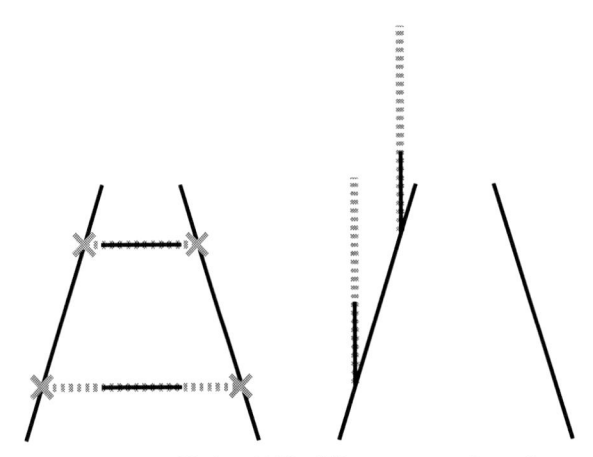

Figure 5.6 Explanation of the Ponzo illusion. A) The different amounts of space between the two identical target lines and the convergent "inducing" lines cause the maximum length (gray dashed lines) for the target near the convergent end to be less than this length near the opposite end. B) The maximum target length is not restricted by contextual lines in this variant of the standard Ponzo stimulus in which the illusory effect is diminished.

Of course, the explanation of size contrast offered is not limited to circles; the same argument applies to the apparent size of any geometrical form in some sort of spatial context. For instance, a superficially quite different size contrast effect is the so-called Ponzo illusion (see Figure 1.2C and Figure 5.6). In the Ponzo stimulus, two identical horizontal lines look different in length in the context of two converging lines, the line closer the point of convergence of the contextual lines appearing longer. This perceptual effect can be explained in much the same way as the different apparent sizes of the inner circles in the Delboeuf figure. As in the case of a Delboeuf stimulus in which the size of the outer circle limits the maximal size of the inner target circle, the converging context lines in the Ponzo stimulus limit the maximum possible length of the horizontal target lines. The maximum length of a target near the convergent end of the contextual lines is thus shorter than the maximum target length near the opposite end (Figure 5.6A). This difference causes the length of a given line to rank higher on the empirical scale of target length near the convergent end compared to the ranking of the same length on the scale for targets nearer the opposite end. Accordingly, the same line near the convergent end of the contextual lines in the Ponzo stimulus should look longer than the other target line, as it does.

Recall that when the horizontal lines in the standard Ponzo stimulus are rotated such that they become vertical attachments to one of the contextual lines, they no longer appear very different in length (see Figure 1.4B). The lack of "illusory" effect in this instance is also readily understood in the present framework. The range of possible target lengths for the vertical attachments in this variant of the Ponzo stimulus is not restricted by the contextual lines, since the targets can extend upward indefinitely (Figure 5.6B). Thus the empirical scale of target length near the convergent end will not differ much from the empirical scale associated with the opposite end. Accordingly, a vertical target near the convergent end of the stimulus would have about the same empirical rank as the identical target near the opposite end, meaning that the targets should appear about the same length, as they do.

This argument, however, should not obscure the fact that multiple empirical factors determine the relevant probability distributions for the perception of any of these stimuli. In the Ponzo stimulus, for example, the space between the target lines and the convergent lines is only a major factor in the effect; other categories of information, such as the length of the converging lines or perspective, are also relevant factors in the complete statistics that underlie these perceptions.

COMPARISON OF AN INNER CIRCLE WITH A SINGLE CIRCLE

When the inner circle of a Delboeuf figure is compared to a single circle of identical size, the inner circle looks larger than the single circle if the outer circle has less than twice the diameter of the inner circle (see Figure 5.1D). Conversely, it appears smaller than the single circle when the diameter of the outer circle is more than 4 or 5 times that of the inner circle (see Figure 5.1E).

To examine whether these seemingly paradoxical phenomena can also be explained by the statistical relationship of images and their sources, consider first the probability distribution of the physical sources of a single circle in the absence of a contextual circle (Figure 5.7A). For the reasons already given, the probability of encountering the real-world source of a projected circle decreases as the projected size of the circle increases. Now compare the probability distribution for a single circle to the distribution of the physical sources of inner target circles in the presence of an outer contextual circle that is either relatively small (Figure 5.7B) or relatively large (Figure 5.7C). Figure 5.7B shows the position of a target 24 pixels in diameter (dashed line) in relation to the probability distribution of the sources of single circles, and in relation to the distribution of the sources of inner target circles given the presence of an outer circle 32 pixels in diameter. The cumulative probabilities associated with the target 24 pixels in diameter derived from these two distributions (Figure 5.7B, inset) show that the same target circle ranks higher on the empirical scale of target size associated with the outer circle than on the empirical scale for single circles. This relationship predicts that when the ratio of the diameter of the outer circle to the target circle is less than 2 (it is 1.33 in the example shown here), the target within the outer circle should appear larger than a single circle of the same size, as it does.

When, however, the contextual circle is significantly larger, this relationship no longer holds. Figure 5.7C shows the position of the same target circle (dashed line) in the probability distribution of the sources of single circles and in the distribution of the sources of inner target circles in the presence of a much larger outer circle (five times the diameter of the target). The cumulative probabilities in this case indicate that the rank of the target is now lower in the context of the outer circle than on the empirical scale for single circles (Figure 5.7C, inset). This further statistical relationship predicts that when the ratio of the diameter of the contextual circle to the target circle

Size

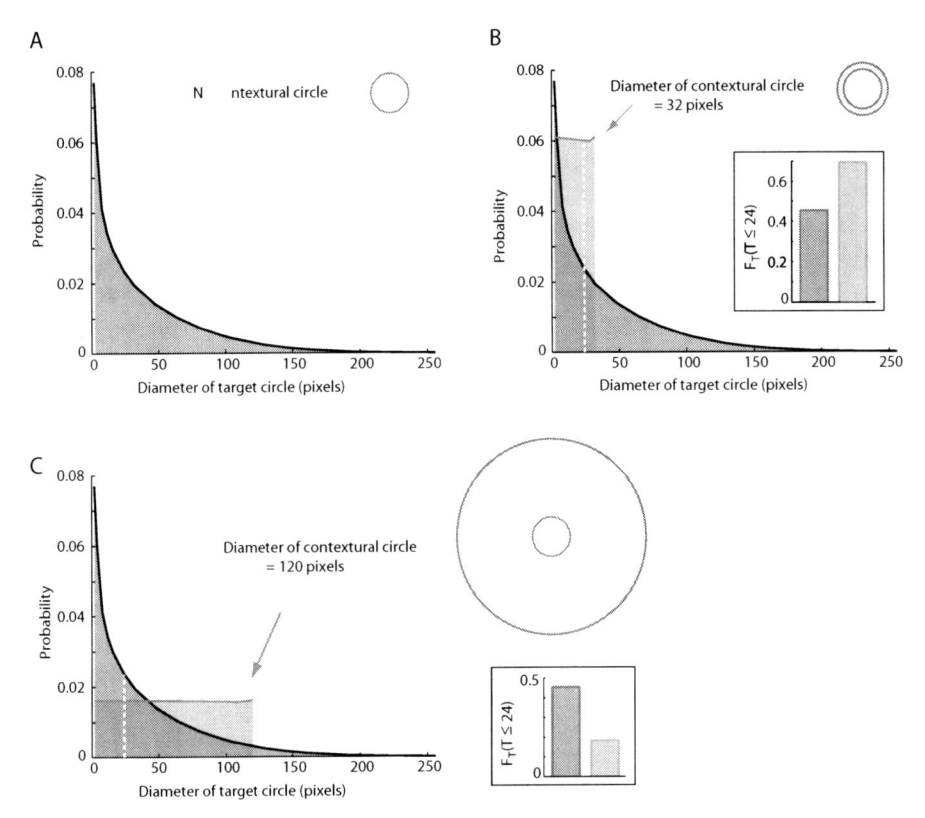

Figure 5.7 Comparison of the inner circle of a Delboeuf figure to a single circle of the same size. A) The probability of occurrence of the physical sources of a single circle as a function of the diameter of the circle in the image plane. B) The probability distribution of the sources of single circles in (A) superimposed on the probability distribution of the sources of inner (target) circles, given the presence of an outer circle 32 pixels in diameter. The dashed line indicates a target circle 24 pixels in diameter as an example. C) The probability distribution of the sources of single circles superimposed on the probability distribution of the sources of inner circles, given the presence of an outer circle 120 pixels in diameter. Dashed line again indicates a target 24 pixels in diameter. Insets in (B) and (C) show the cumulative probabilities $F_T(T \leq 24)$ derived from the corresponding probability distributions. (After Howe and Purves, 2004)

is large, the target in the contextual circle should look smaller than an identical single target circle. Again, this is what is seen.

THE POINT OF PERCEPTUAL TRANSITION

Several psychophysical studies have varied the ratio of the diameter of the outer circle to the inner circle in the Delboeuf stimulus in attempts to determine the exact point at which the perceived size of the inner circle, compared to a single circle, changes from overestimation to underestimation (cf. Figures 5.1D and E) (Ikeda and Obonai, 1955;

Oyama, 1960). Oddly, this ratio varies according to the absolute size of the target circle, decreasing as the visual angle subtended by the target increases. This fact presents yet another challenge to any explanation of size contrast and assimilation.

In the empirical framework here, the point of perceptual transition should correspond to the ratio of the outer circle to the inner target circle at which the target has the same empirical rank in the context of the outer circle as when presented alone. Consider a target circle x pixels in diameter, and let F_{T0} denote the cumulative probability $F_T(T \leq x)$ derived from the probability distribution of the sources of single circles, and F_{T1} the cumulative probability derived from the distribution of the sources of target circles given an outer contextual circle. As already shown in Figure 5.7, F_{T1} decreases as the diameter of the outer circle increases (compare Figure 5.7B with C), becoming equal to F_{T0} when the diameter of the outer circle equals some particular value (D_0 in Figure 5.8). Accordingly, when the diameter of the outer circle reaches this value, a target circle x pixels in diameter within the outer circle should appear equal in size to an identical single circle. Therefore, the ratio of the diameter of the outer circle to the target circle where the perceptual transition of relative size occurs should be equal to D_0/x. As is apparent in Figure 5.8, this ratio decreases as x increases. When x equals 24 pixels (equivalent to a visual angle of $\sim 3.5°$), the ratio D_0/x is about 2.2 (Figure 5.8A); in contrast, when x is 32 pixels (a visual angle of $\sim 4.6°$), the ratio is 1.8 (Figure 5.8B). Thus, the variable point of perceptual transition as a function of the visual angle subtended by the target circles is also predicted by empirical ranking.

COMPARISON OF THE OUTER CIRCLE IN A CONCENTRIC SET WITH A SINGLE CIRCLE

A final aspect of the perceptual effects associated with the Delboeuf figure is that the *outer* circle of a concentric pair appears smaller than an identical single circle, an effect that gradually diminishes as the inner circle of the concentric pair becomes smaller (see Figure 5.1F and G).

Figure 5.9A compares the probability distribution of the sources of single circles with the distribution of the sources of outer target circles, given an inner circle 32 pixels in diameter. This latter distribution shows that the context provided by an inner circle increases the probability of occurrence of the sources of outer circles when the size of the outer circle is not much larger than the size of the inner circle (which is now the contextual circle). This distribution, however, converges with the distribution of the sources of single circles as the outer circle becomes increasingly large. A target circle 48 pixels in diameter, for example (dashed line in Figure 5.9A), obviously has different positions in these two probability distributions. The cumulative probabilities $F_T(T \leq 48)$ derived from the two probability distributions indicate that the empirical rank of the size of this target is lower in the context of an inner circle of 32 pixels than in relation to the size of single circles (see inset in Figure 5.9B). These statistics predict that the outer (target) circle presented with an inner (contextual) circle should appear smaller than a single circle of the same size.

These same statistics also predict the perceptual responses observed as the inner circle becomes increasingly smaller. In addition to the two probability distributions

Size

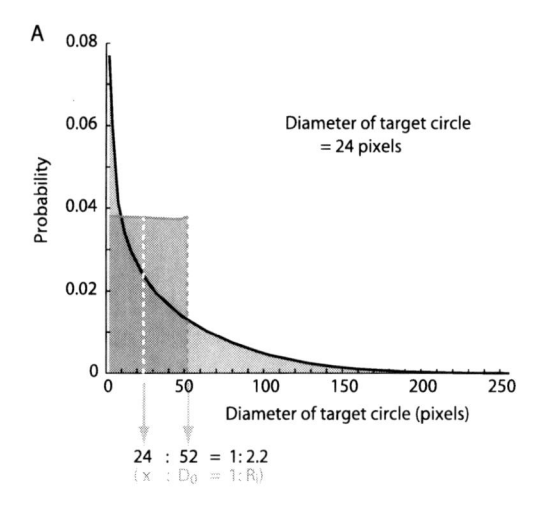

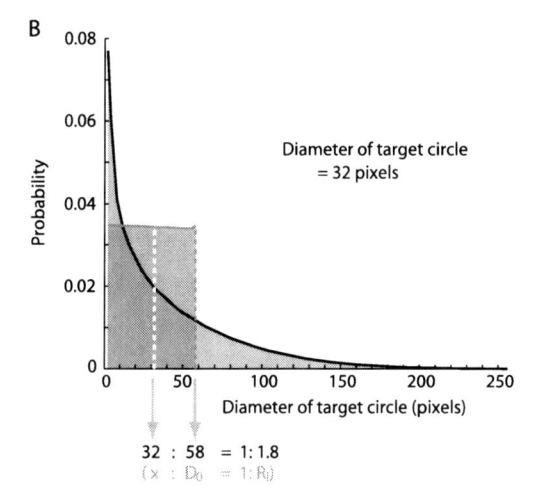

[Figure 5.8 - Chart A: Probability distribution with target circle diameter = 24 pixels, showing 24 : 52 = 1:2.2 ($x : D_0 = 1 : R_i$)]

[Figure 5.8 - Chart B: Probability distribution with target circle diameter = 32 pixels, showing 32 : 58 = 1:1.8 ($x : D_0 = 1 : R_i$)]

Figure 5.8 Predicting the ratio of the diameter of the outer circle to the inner circle in the Delboeuf stimulus at which the apparent size of the inner circle equals that of an identical single circle. A) The probability distribution of the physical sources of single circles (black) is superimposed on the probability distribution of the sources of inner (target) circles, given an outer (contextual) circle of diameter D_0. In this example, in which the target circle is 24 pixels in diameter, D_0 must be 52 pixels for the cumulative probabilities of this target derived from the two distributions to be equal, predicting a ratio (R_i) at perceptual equality of 2.2. B) When the diameter of the target circle is 32 pixels, D_0 must be 58 pixels for the cumulative probabilities $F_T(T \leq 32)$ to be equal, predicting a ratio of 1.8. Thus the ratio of the diameters of the outer circle to the inner circle at the perceptual transition decreases as the absolute size (i.e., the visual angle) of the target circle increases, in accord with psychophysical reports. (After Howe and Purves, 2004)

in Figure 5.9A, Figure 5.9B shows the distributions of the sources of outer target circles, given an inner circle 16 or 8 pixels in diameter. The cumulative probabilities derived from these distributions increase as the size of the inner circle decreases (see Figure 5.9B inset). In fact, when the inner circle is very small relative to the outer circle (e.g., 8 vs. 48 pixels in diameter), the cumulative probabilities indicate that the

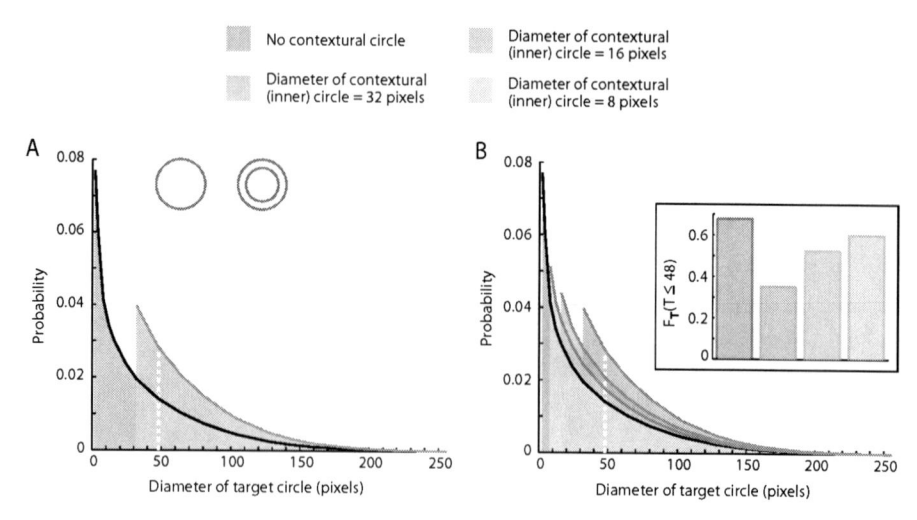

Figure 5.9 Comparison of the outer circle of a Delboeuf figure to a single circle. A) The probability distribution of the sources of single circles (black; see Figure 5.7A) superimposed on the probability distribution of the physical sources of outer (target) circles, given the presence of an inner (contextual) circle of 32 pixels in diameter. The dashed line indicates a target circle 48 pixels in diameter, as an example. B) The probability distribution of the sources of single circles superimposed on the probability distributions of the sources of outer circles, given the presence of an inner circle 32, 16 and 8 pixels in diameter, respectively. Inset shows the cumulative probabilities for an outer target circle 48 pixels in diameter derived from these four distributions. (After Howe and Purves, 2004)

outer target circle in the context of the inner circle should appear about the same size as an identical single circle, as indeed it does (see Figure 5.1G).

OTHER THEORIES OF SIZE CONTRAST AND ASSIMILATION

Many ideas have been put forward to explain size contrast and assimilation effects during the century or more that these phenomena have been a topic of serious research. The explanation of size contrast most often cited is the "adaptation level" theory, which suggests that the overall "level" of a stimulus ("size level" in this instance) provides a reference or "anchor point" that the visual system then uses to make judgments about the relative magnitude of specific stimulus elements (Helson, 1964; Green and Stacey, 1966; Restle and Merryman, 1968; Jordan and Uhlarik, 1986). For a size contrast stimulus such as the Ebbinghaus figure, the different "levels" of adaptation generated by the different sizes of the surrounding circles would, in this conception, make the two identical central targets appear different in size. This argument, however, cannot explain the effect of altering the interval between the central and surrounding circles of the Ebbinghaus stimulus (see Figure 5.1B) or many of the other phenomena elicited by the stimuli in Figure 5.1. In particular, the theory is contradicted by size assimilation effects. For example, adaptation level theory predicts that the inner target circle in Figure 5.1D should appear smaller than the single circle because the overall size of the concentric circles is larger, thus providing a higher "adaptation level" than the single circle. This prediction is opposite the perception elicited. A number of other approaches that seek to rationalize aspects of size contrast in this general way suffer from the same deficiency (see, for example, Ganz, 1966; Jaeger and Pollack, 1977; Jaeger and Lorden, 1980).

An approach that purports to rationalize both size contrast and assimilation has been to imagine that the words "contrast" and "assimilation" signify in a more literal sense the cognitive operations that perform these comparisons. In this conception, size contrast effects are taken to be the consequence of comparing the properties of an object with its context, whereas assimilation is considered to be the result of incorporating the properties of contextual elements into the percept of the target (Coren, 1971; Coren and Enns, 1993; Rock, 1995). A problem with this way of thinking is the lack of any biological reason why such processes would be necessary, or even useful. Moreover, it is not clear under what conditions the mechanisms of contrast versus assimilation would or should operate, or how contrast and assimilation could be investigated in these terms.

SUMMARY

The diversity of size contrast and assimilation effects reviewed in this chapter has generally defied attempts at coherent explanation. The analyses and observations summarized here show that, like simpler stimuli that involve single lines, angles formed by two lines or combinations of lines that cause tilt effects, the variety of size contrast and assimilation effects that have been described over the years are well accounted for in terms of the probability distributions of the possible sources of the relevant stimuli.

Chapter 6

Distance

The way we see lines, angles and the relative size of objects are all aspects of the way human beings perceive physical space. Because—as should be clear from the variety of anomalies described in the preceding chapters—this subjective experience does not correspond in any simple way to physical space, perceived space is referred to as "visual space". For many who have thought about this issue, an appealing intuition has been that the properties of visual space arise as the result of a transformation of the Euclidean properties of physical space (Indow, 1991; Hershenson, 1999). However, it is obviously difficult to reconcile the predicted consequences of any direct transform of physical space with the many peculiarities of what people actually see; a direct transform would imply that anything that is longer, larger or further away in physical space should be seen as such, but this is clearly not the case. Indeed, the anomalies already discussed provide critical clues to understanding how visual space is actually generated.

Based on the specific perceptions of geometry covered in earlier chapters, these clues suggest that the way we perceive space generally arises from the statistical manner in which observers link visual stimuli to the real-world sources that human beings have encountered over the course of species and individual experience. A fundamental descriptor of visual space not yet considered—and certainly one that is pertinent to this concept—is the perception of egocentric distance (i.e., the apparent distance from the observer to objects in the world; the account that follows is based on detailed observations presented in Yang and Purves, 2003). The purpose of this chapter is thus to consider whether the perception of this further aspect of space can also be explained in terms of the statistical relationship between images and their physical sources.

CHARACTERISTICS OF APPARENT DISTANCE

The phenomenology of distance perception clearly provides some puzzles that need to be explained. As illustrated in Figure 6.1, it has long been known that the apparent

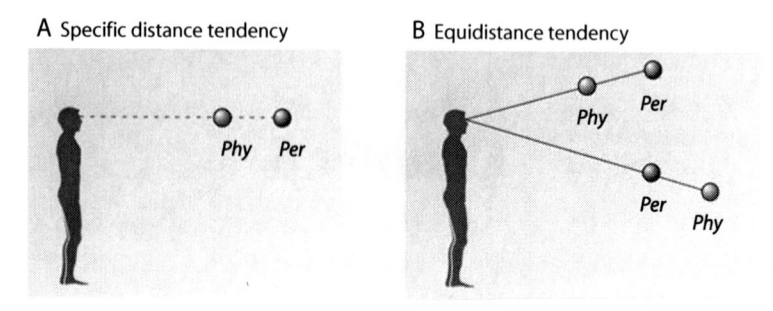

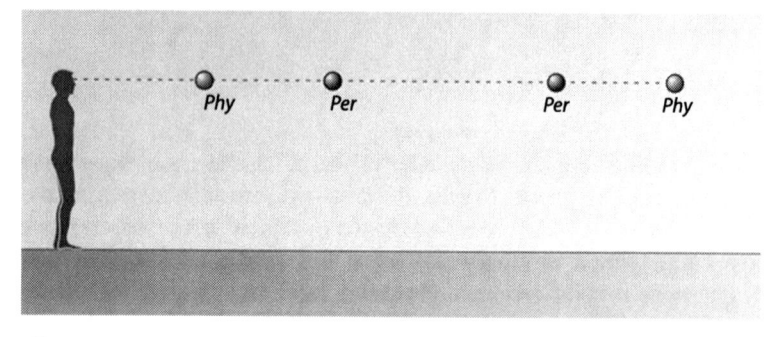

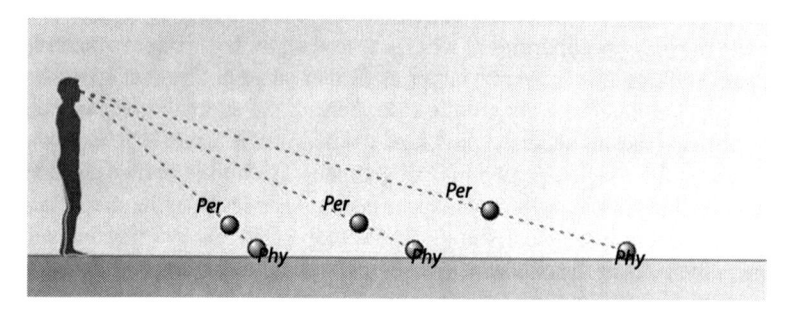

Figure 6.1 Anomalies in perceived distance, each of which is explained in the text. In these diagrams, which are not to scale, "Phy" indicates the physical position of the object, and "Per" the perceived position. (A) The specific distance tendency. (B) The equidistance tendency. (C) The perceived distance of objects at eye-level. (D) The perceived distance of objects on the ground. (After Yang and Purves, 2003)

distance of objects bears a peculiar relationship to their physical distance from the observer (Sedgwick, 1986; Gillam, 1995; Loomis et al., 1996; Hershenson, 1999).

When subjects are asked to make judgments with little or no contextual information (e.g., the distance of a luminous but otherwise featureless object in a darkened room), the distances reported differ in several ways from the corresponding physical distances. First, objects in these circumstances are typically perceived to be at a distance of 2–4m, a phenomenon referred to as the "specific distance tendency" (Gogel, 1965;

Distance

Owens and Leibowitz, 1976) (Figure 6.1A). Second, objects that are relatively near each other in the retinal image appear to be about the same distance from the observer, a phenomenon called the "equidistance tendency" (Gogel, 1965) (Figure 6.1B). Third, when presented at or near eye-level, the distance of objects relatively near the observer tends to be overestimated, whereas the distance of objects that are further away tends to be underestimated (Epstein and Landauer, 1969; Gogel and Tietz, 1979; Morrison and Whiteside, 1984; Foley, 1985; Philbeck and Loomis, 1997) (Figure 6.1C). Finally, the apparent distance of objects on the ground varies with the angle of declination of the line of sight (Wallach and O'Leary, 1982): objects on the ground that are at least several meters away appear closer than they really are, and with increasing distance are judged to be progressively more elevated than warranted by their physical position (Ooi et al., 2001) (Figure 6.1D).

Although a variety of explanations have been proposed in the various studies cited, there has been little or no agreement about the basis of these unusual perceptions of egocentric distance. More often than not, the several tendencies illustrated in Figure 6.1 have simply been accepted as empirical facts that are then used to rationalize other aspects of visual space.

Given the ability of the probabilistic relationship between retinal images and sources to explain a variety of other geometrical percepts, it makes sense to ask whether the probability distributions of the possible sources of visual stimuli also determine apparent distance. Accordingly, the database of natural scene geometry described in Chapter 2 was used to test whether the anomalous perceptions of distance illustrated in Figure 6.1 can be accounted for by the probability distributions of the physical distances of object surfaces from human observers.

PROBABILITY DISTRIBUTIONS OF PHYSICAL DISTANCES IN NATURAL SCENES

The first step in this assessment was to use the distance information for each pixel in the range images to compute the distribution of the distances from the image plane of the scanner (a proxy for the retinal image plane of an observer) to all the locations in the physical scenes in the database. In analyzing this information in what follows, a number of relatively special terms are required, including ground plane, radial distance, elevation angle and others; these are defined graphically in Figure 6.2.

The first of several statistical features apparent in this sort of analysis is that the probability distribution of the radial distances from the image plane to physical locations in the scenes has a maximum at about 3m, declining approximately exponentially over greater distances (Figure 6.3A). This decay is presumably due to the fact that the farther away an object is, the less the area it spans in the image plane and the more likely it is to be occluded by other intervening objects. The falloff at very near distances arises in large part because the range scanner was never placed directly in front of large objects that would have prevented the beam from scanning the majority of a scene (see Chapter 2); as a result there is very little probability mass at distances < 1m. Although as noted earlier this bias is largely an artifact of the way we sampled the environment, it is

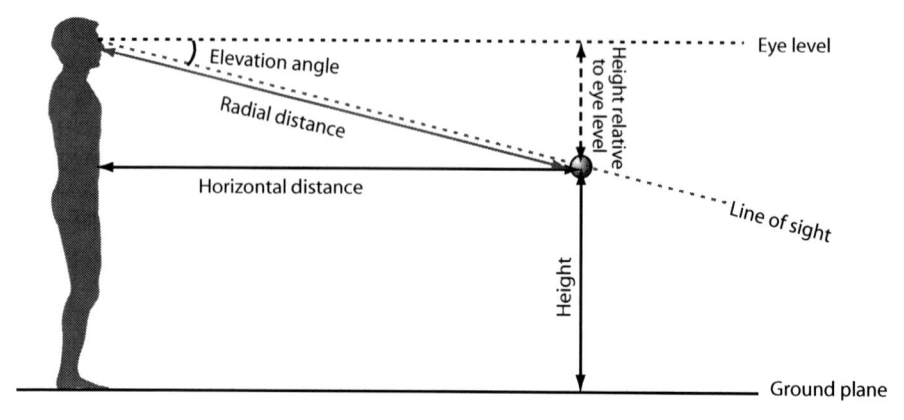

Figure 6.2 Diagram defining the various terms used to describe scene geometry with respect to an observer.

reasonable to presume that human observers in natural settings generally behave in this same way (i.e., avoid nearby objects that block vision) to better see their surroundings.

A second feature emerging from the analysis concerns how different physical locations in natural scenes are typically related to each other in terms of their radial distance from the observer. The distribution of the absolute differences in the radial distance from the image plane of the scanner to any two physical locations (whether separated horizontally or vertically) is highly skewed, having a maximum near zero and a long tail (Figure 6.3B; the maximum difference is very close to 0 for horizontally separated locations, and at about 10–15cm for vertically separated locations). Even for physical separations as large as 30° of visual angle, the most probable difference between the distances from the image plane to the two locations is relatively slight, a result that is not at all obvious from simply inspecting the world around us.

A third statistical feature that emerges from the range data is that the probability distribution of horizontal distances from the scanner to physical locations (i.e., distances from the scanner in the horizontal plane; see Figure 6.2) changes relatively little with the height of locations in the scene (recall that the height of scanner was always adjusted to be 1.65m above the ground, thus approximating the height of the average adult viewpoint, referred to here as "eye level") (Figure 6.3C). The probability distribution of physical distances at eye level has a maximum at about 4m and decays gradually over greater distances. The distributions of the horizontal distances of surface locations at different heights above and below eye level have roughly the same shape as the distribution at eye level.

SIGNIFICANCE OF THESE STATISTICS FOR PERCEIVED DISTANCE

The question of interest, then, is whether these probability distributions of distances from the image plane in natural scenes can account for the anomalies of apparent distance summarized in Figure 6.1. In terms of the empirical framework used to

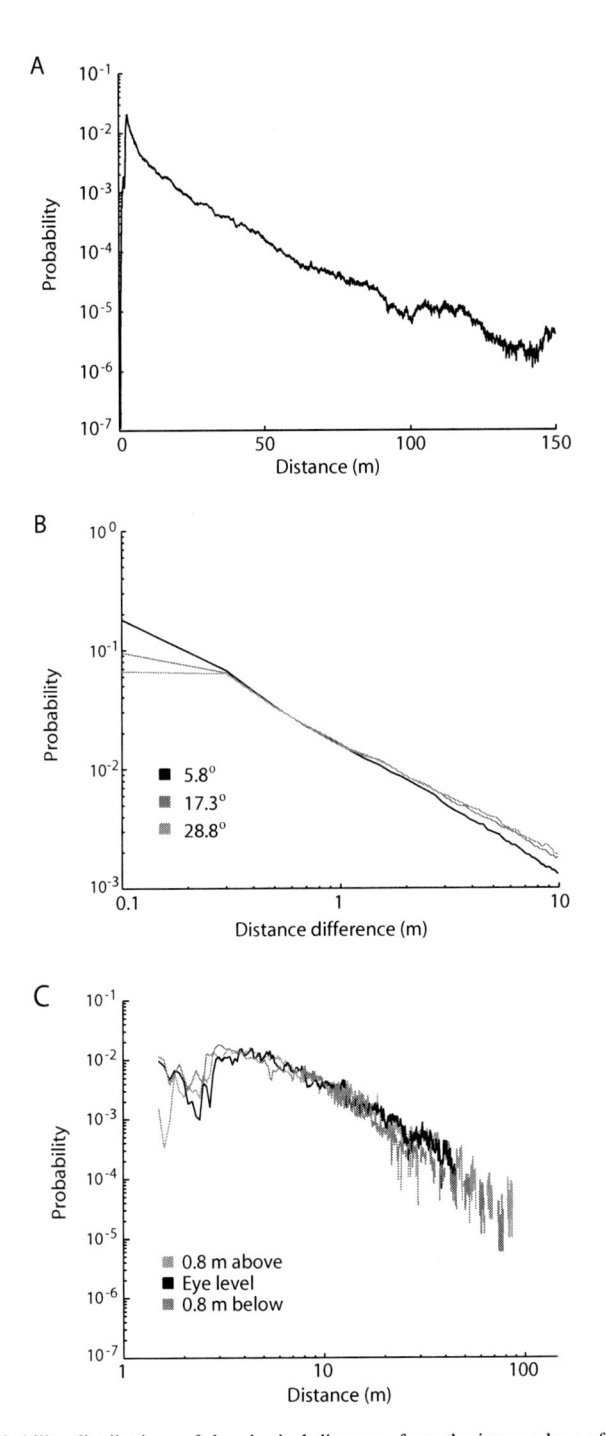

Figure 6.3 Probability distributions of the physical distances from the image plane of the laser range scanner to the surfaces of objects in the scenes in the database. (A) The probability distribution of distances from the center of the scanner to all surface locations. (B) The probability distribution of the differences in the distances of two surface locations in physical space separated by three different visual angles in the horizontal plane (vertical separations, which are not shown here, had a generally similar distribution). (C) Probability distributions of the horizontal distances of surface locations at different heights with respect to eye level (see Figure 6.2). (After Yang and Purves, 2003)

rationalize the perceptual anomalies discussed in previous chapters, the probability distributions of distances illustrated in Figure 6.3 provide a basis for constructing a subjective "space" of apparent distances according to how frequently human observers have encountered different physical distances in the past. The apparent distance of any point in a visual stimulus would thus be determined by the influence of this accumulated experience.

The Specific Distance Tendency

As described, the sort of stimuli used to demonstrate the so-called specific distance tendency and other anomalies in the perception of distance have typically entailed a featureless target in darkness with no other visual information (e.g., a dim source of light). Thus, unlike stimuli examined in previous chapters where a measurable feature of the stimulus (e.g., the length of a line, the subtense of an angle, the size of an object) could be evaluated against the relevant empirical scale of that feature, the stimuli used to assess egocentric distance contain essentially no geometrical information other than the location of a luminous blob (for instance) in a darkened visual field. Given these circumstances, the perceived distance should simply accord with the distances most frequently encountered in the past experience of human observers.

As indicated in Figure 6.3A, the overall probability distribution of the distances of object surfaces in natural scenes from observers has a strong maximum at about 3m. In other words, there are far more points in natural scenes associated with this range than with other distances. This distribution thus predicts that, in the absence of specific geometrical information in a visual stimulus, the apparent distance of objects in the subjective "space" that reflects this aspect of geometrical perception will be about 3m. This prediction is in agreement with the evidence reported in psychophysical studies that observers tend to perceive objects to be at a distance of 2–4m under these experimental conditions (see Figure 6.1A).

The Equidistance Tendency

The similar distance of neighboring points perceived in the absence of additional information (see Figure 6.1B) also accords with the probability distributions of the distances to surface locations in natural scenes. Since the probability distribution of the differences of the physical distances from the image plane to two locations with relatively small angular separations (the black line in Figure 6.3B) has a maximum near zero, any two neighboring objects should, if apparent distance is determined empirically, be perceived to be at about the same distance from the observer in the absence of additional information in the visual stimulus. At large angular separations (the colored lines in Figure 6.3B), however, the probability associated with small absolute differences in the distance to the two points is somewhat smaller, and the distribution progressively flatter. Accordingly, the tendency to see neighboring points at the same distance from the observer would be expected to decrease somewhat as a function of increasing angular separation. This tendency has also been observed in psychophysical studies, although it has not been documented quantitatively (Gogel, 1965). The agreement between this psychophysical evidence and the distribution of relative distances as a function of

Distance

object separation is thus consistent with a probabilistic explanation of the equidistance tendency.

Perceived Distance of Objects at Eye Level

The probability distribution of physical distances at eye level (the black line in Figure 6.3C) can, in much the same way, account for the perceptual anomalies in response to stimuli generated by near and far objects presented at this height (see Figure 6.1C). Based on this probability distribution, the perceived distance of an object located at eye level should, in the absence of other contextual information in the retinal image, be perceived to be about 4m away. Therefore, the distance of an object that is actually located closer than 4m would be overestimated and the distance of an object farther than 4m would be underestimated.

These predictions again fit the psychophysical data quite well. For instance, Philbeck and Loomis (1997) showed that the apparent distance of a dimly luminous object presented at eye level in an otherwise dark environment tends to remain at about 4m as the actual distance is varied between 2 and 5m (subjects reported apparent distance in this case by walking blindfolded to the place they thought the object was, explaining the relatively small range of distances tested).

The Apparent Distance of Objects on the Ground

To examine whether the perceptual phenomena illustrated in Figure 6.1D can also be accounted for in these terms, the probability distribution of physical distances of points at different elevation angles relative to the horizontal plane at eye level (i.e., along different lines of sight; see Figure 6.2) was also determined by analyzing the range database.

As shown in Figure 6.4A, the probability distribution of physical distances is more dispersed when the line of sight is directed above rather than below eye level. Moreover, the distribution shifts toward nearer distances as the line of sight deviates increasingly from eye level, a tendency that is more pronounced below than above eye-level.

These statistical differences as a function of the elevation of the line of sight are more obvious in Figure 6.4B, which shows the most likely physical distances (solid black line) to locations below eye level in the scene database as a function of the elevation angle. The most likely distances from the eyes of an observer to surface locations at various elevations form a curve that is relatively near the ground at closer horizontal distances, but increases in height as the horizontal distance from the observer increases. The physical basis of this variation is presumably that objects below eye level that are progressively farther from the observer have to be increasingly higher in the visual field to be visible; more distant objects that are low lying tend to be occluded by objects closer to the observer.

These further statistical characteristics of physical distances can thus account for the perceptual effects illustrated in Figure 6.1D, i.e., that the perceived location of an object on the ground in an featureless environment is influenced by the declination of the line of sight, the object appearing closer and higher than it really is as a function

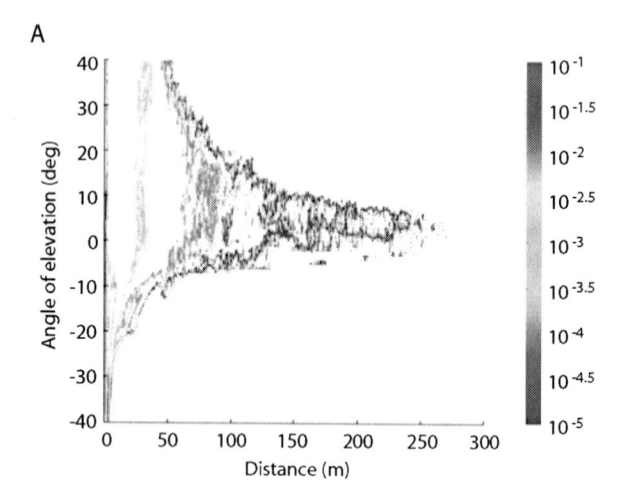

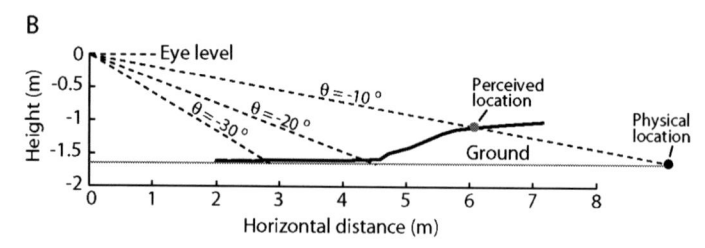

Figure 6.4 Probability distributions of physical distances at different elevation angles. A) Contour plot of the probability distributions of distances at different elevation angles. Probabilities are indicated by color-coding (bar on the right). An elevation angle of 0° is eye level; positive elevation angles correspond to lines of sight above eye level and negative values to those below eye level. B) The most likely radial distance from the image plane as a function of the elevation angle (θ) of the location, derived from the data in (A). The vertical axis is the height relative to eye level; the horizontal axis is horizontal distance from the image plane. The blue line indicates the position of the ground plane at 1.65m below eye level. These statistics predict that the apparent distance of an object on the ground more than a few meters away in a darkened environment should appear closer and higher than the actual location of the object.

of this angle (Wallach and O'Leary. 1982; Philbeck and Loomis, 1997; Ooi et al., 2001). As shown in Figure 6.4B, the distances most frequently experienced by human observers at various elevation angles predict that the apparent location of an object on the ground will indeed be increasingly higher and closer relative to its physical location as the line of sight becomes closer to eye level.

OTHER APPROACHES TO RATIONALIZING VISUAL SPACE

Some earlier studies have variously proposed that visual space represents a Riemann space with constant curvature, an affine space, or a transformation of Euclidean space (Luneberg, 1947; Wagner, 1985; Indow, 1991; Todd et al., 2001). Others have suggested that visual space is a computed composite, based on more or less independent information derived from cues such as perspective, texture gradients, binocular disparity, and motion parallax (Gillam, 1995; Loomis et al., 1996; Hershenson, 1999). The most influential theory, however, was put forward several decades ago by James Gibson (1950; 1979). Gibson argued that since human beings are terrestrial, the ground is the key factor in determining the perception of geometry. In this conception, a 2-D frame of reference built on the Earth's surface is taken to be the basis of visual space. Although this "ecologically" based idea is perhaps closer to the statistical framework presented here, Gibson never specifically dealt with the anomalies of egocentric distance considered in this chapter, or how they might be explained.

Thus whereas each of these approaches has some merit, in terms of the present argument they are in varying degrees off the mark. If visual space is generated by the statistical relationship between images and their sources, then explaining the relevant perceptual phenomenology will inevitably depend on the statistical properties of natural visual environments with respect to visual observers. Absent the empirical information about image-source relationships derived from a range image database, any explanation of visual space is bound to be inadequate, as evidenced by the fact that none of these previous suggestions about visual space has been able to rationalize the full spectrum of discrepancies between physical and apparent distance.

SUMMARY

The ability to explain a variety of subtle anomalies in the perception of distance based on the statistics of the physical distances of object surfaces from the observer in natural scenes offers further evidence that rationalizing perceived geometry in the probabilistic framework outlined in previous chapters is a powerful way of understanding visual space. In addition to successfully explaining a series of specific anomalies that have been difficult to rationalize in other ways, the findings summarized here imply that the sense of egocentric distance is another manifestation of the probabilistic strategy that allows the human visual system to contend with the inherent ambiguity of visual stimuli.

Chapter 7

The Müller-Lyer Illusion

The perception of line lengths, angles, sizes and distances presents a series of relatively straightforward problems in the sense that the challenge is easy to state in geometrical and statistical terms; that is, how we see these elemental aspects of geometry and why. For stimuli that generate the best known of the classical geometrical illusions (see, for instance, Figures 1.2D and E), however, it is much more difficult to state, or even imagine, what the empirical significance of the stimulus really is. Indeed, this difficulty is one of the reasons these effects have been the source of seemingly endless debate and controversy. The examples taken up in this and the following chapters are the infamous Müller-Lyer illusion (Figures 1.2D and 7.1) and the equally perplexing Poggendorff illusion (Figures 1.2E and 8.1). These stimuli and their variants emphasize that to understand the perceptual effects elicited by even relatively simple geometrical stimuli requires going beyond intuitions; the explanations that emerge from the analysis of the database are, as it turns out, quite subtle.

THE MÜLLER-LYER ILLUSION

The Müller-Lyer effect is the arguably best known of the classical geometrical illusions, having been the subject of hundreds of studies since its introduction in the late 19th C. (Müller-Lyer, 1889). The perceptual effect is that two identical straight lines appear different in length when they are terminated, respectively, with "arrowheads" that extend inward with respect to the "shaft", or "arrow tails" that extend outward. As is apparent in Figure 7.1A (and Figure 1.2D), the line terminated by the arrowheads appears shorter than the same line terminated by arrow tails (see Lewis, 1909; Nakagawa, 1958; Dewar, 1967; Gregory, 1968; Earlebacher and Sekuler, 1969 for quantitative studies; there is some variation in the magnitude of the measured effects in these accounts, presumably due to the different experimental conditions and the stimuli used). Thus, in comparison to a line lacking these adornments, the length of the line with the arrowheads is underestimated, and the line with arrow tails overestimated, typically by about 5–10%.

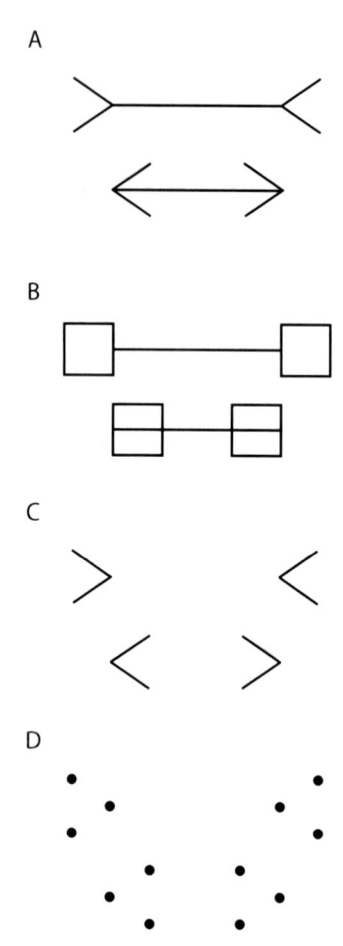

Figure 7.1 The Müller-Lyer effect. A) The standard Müller-Lyer stimulus. Although the central shafts of the two figures are identical, the line terminated by arrow tails appears longer than the line terminated by arrowheads. B) A variation of the Müller-Lyer stimulus in which the arrowheads and tails are replaced by squares. Despite this substitution, the illusory effect remains much the same. C) A similar effect is generated by a variant in which the central shafts are missing. D) The Müller-Lyer effect can also be elicited, although less strongly, by a figure comprising only a few dots. (After Howe and Purves, 2005b)

Rationalizing this illusion has been made especially difficult by the persistence of the effect when the identical lines are terminated with a variety of other adornments, a fact that was noted by Delboeuf (1892) not long after Müller-Lyer's original description. This peculiarity undermines intuitive explanations of the effect based on what the specific geometries of arrowheads and tails might signify to an observer. In Figure 7.1B, for instance, much the same effect is generated when the lines are terminated by outward and inward squares. Further obstacles to any simple explanation of the Müller-Lyer effect are that neither the presence of the shaft (Figure 7.1C) nor even lines (Figure 7.1D) are needed to elicit a misperception of the relevant spatial interval. As a result of these several observations, there has been a great deal of controversy about the genesis of the Müller-Lyer effect (e.g., Nakagawa, 1958; Gregory, 1963, 1966; Pressey, 1967; Bolles, 1969; Day, 1972; Griggs, 1974; Morrison, 1977; Bross et al., 1978; Skottun, 2000), which still has no generally accepted explanation (Rock, 1995; Robinson, 1998).

SAMPLING THE PHYSICAL SOURCES OF MÜLLER-LYER STIMULI

The Müller-Lyer stimulus thus presents an obvious and difficult test of the statistical framework that has been successful in rationalizing the perceptual anomalies generated by the simpler stimuli already considered. In keeping with the statistical approach applied to these other geometrical illusions, both the standard presentation of the Müller-Lyer stimulus in Figure 7.1A and the variants shown in Figure 7.1B-D were used to test the hypothesis that the identical lines or intervals in these stimuli appear different in length because the probability distributions of the real-world sources of the lines or intervals, given the different contexts provided by the adornments in the Müller-Lyer stimulus, are in fact different.

An intuitively appealing way to sample the image database for possible physical sources of the Müller-Lyer configuration—or any geometrical figure—would be to identify those areas of the images that contain a pattern of luminance contrasts (i.e., edges) corresponding to the Müller-Lyer stimulus. As discussed in Chapter 3, however, this approach is neither feasible nor conceptually appropriate. From the perspective of feasibility, luminance patterns in the form of the Müller-Lyer stimulus almost never occur in natural scenes, and are rare even in indoor scenes replete with "carpentered" objects (see below). Thus, an analysis of the database carried out in this way would yield only a small number of samples that would have commensurately little statistical meaning. Equally important, from a conceptual standpoint the percepts associated with the Müller-Lyer (or any geometrical) stimulus presumably derive from experience with *all* the geometrical arrangements relevant to generating appropriate behavioral responses to the stimulus under consideration (see Chapters 2 and 3; indeed the virtual absence of explicit Müller-Lyer stimuli based on contrast boundaries in nature strongly implies this). Accordingly, all the points in the database whose geometry conformed to the Müller-Lyer configuration were considered valid samples.

Determination of the physical sources of the Müller-Lyer stimulus and the subsequent computation of the relevant probability distributions of these sources involve several steps. First, an appropriate template was applied to the images to identify areas of the scenes that contained physical sources of *one* of the pair of adornments in a Müller-Lyer figure (i.e., an arrowhead, an arrow tail or the equivalent in the Müller-Lyer variants) (Figure 7.2A). As indicated in Figure 7.2B, the set of pixels underlying the template was then screened to determine whether the physical points corresponding to each straight line in the template formed a geometrically straight line in 3-D space (see Chapter 3). If this criterion was met, the points were accepted as a valid sample of the physical source of the *conditional adornment*.

Once a valid physical source of the conditional adornment had been identified, the same region of the scene was then tested for the occurrence of the other components of the Müller-Lyer figure by overlaying on the same image a series of templates

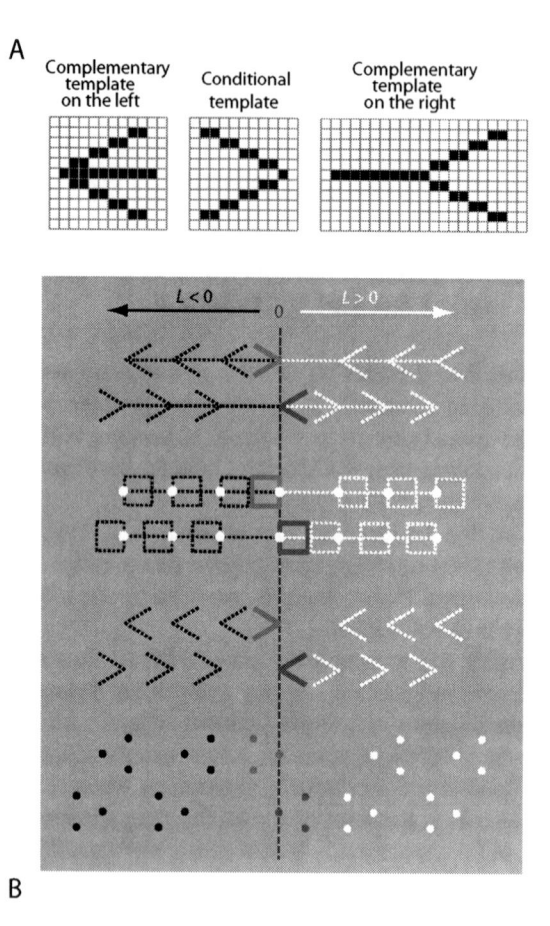

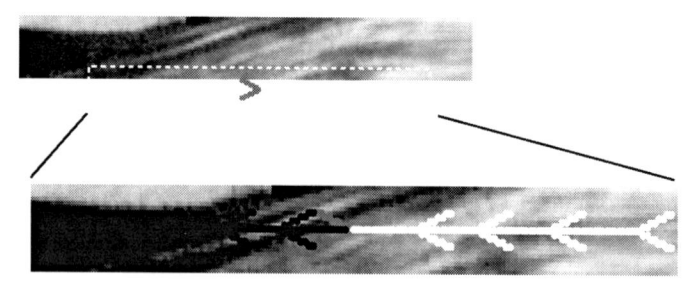

Figure 7.2 Sampling the range image database with templates pertinent to the Müller-Lyer stimulus. A) The region in the image to be analyzed is represented diagrammatically by the grid squares in the upper panel, each of which corresponds to an image pixel; the sets of black pixels indicate examples of templates used to sample different elements of the standard Müller-Lyer figure. Each row in the lower panel illustrates a conditional template (red or blue) in the first step of the sampling procedure, and a series of complementary templates (black or white) subsequently applied to the image (only a few examples in the full series actually tested are shown). The white dots on the squares indicate the reference edge of these adornments. The length (L) of the target (shaft) line or interval was designated positive when the complementary template was on the right side of the conditional adornment, and negative when the complementary template was on the left. B) The sampling procedure applied to an image. The red template indicates a valid sample of the conditional adornment that met the geometrical criteria described in the text; a series of complementary templates was then overlaid at successively greater distances from the conditional adornment to assess the presence of valid samples of the complementary adornment, as indicated in the blow-up. (After Howe and Purves, 2005b)

complementary to the template for the conditional adornment (see Figure 7.2B). For a standard Müller-Lyer figure, these complementary templates comprised a shaft of increasing length and an arrow adornment configured as the mirror image of the conditional adornment being examined. For the Müller-Lyer variant with squares, the complementary templates comprised a square with a shaft of increasing length attached to either the left or the right edge of the square. In the case of the Müller-Lyer variants without a shaft, or comprising only dots, the complementary template was simply a mirror reflection of the conditional template. This second step thus identifies the *complementary adornment* and the shaft or interval between the two adornments.

The length of the shaft (or the corresponding interval) was varied incrementally, negative values indicating a complementary template to the left of the conditional adornment, and positive values to the right. Note that as the complementary template shifts from the left to the right of the conditional adornment, the overall configuration of the stimulus formed by these two components reverses (see Figure 7.2A). As for the conditional adornment, the physical points corresponding to each straight line in the complementary template were evaluated to see whether they also formed a straight line in 3-D space. If this further criterion was met, the sample was counted as a valid physical source of the Müller-Lyer figure in the configuration specified by the combination of the conditional and complementary templates.

By repeating this general procedure for many regions of the images in the database and counting the total number of valid samples of physical sources identified by each different combination of conditional and complementary templates, the probability of occurrence of physical sources of Müller-Lyer projections as a function of the projected length of the shaft (or the interval between the two adornments) was tallied up. We could then ask how the probability distributions obtained in this way vary according to how a shaft or interval is adorned at its ends.

ANALYSIS OF THE STANDARD MÜLLER-LYER STIMULUS

The probability distributions of the physical sources of the standard Müller-Lyer stimulus with varying shaft lengths (L) obtained from images of the fully natural scenes in the database are shown in Figure 7.3A. The distribution indicated in black was derived from a sampling procedure in which the apex of the conditional adornment pointed to the right; the distribution in gray was derived using a conditional adornment whose apex pointed to the left (see diagram in Figure 7.3A). Thus the left half of the distribution in black (where $L < 0$) corresponds to shafts adorned with arrowheads, whereas the right half (where $L > 0$) corresponds to shafts with arrow tails; the opposite is true for the distribution shown in gray. As evident in the figure, there is a systematic difference between these two probability distributions. In relation to the point at which $L = 0$, the mode of the black distribution is shifted to the left, whereas the mode of the gray distribution is shifted to the right. Furthermore, for each value of L less than 0, the distribution represented in black gives a higher probability than the gray distribution, whereas the opposite applies for all values of L larger than 0.

With respect to the perceptual consequences described in the next section, a useful way of summarizing the differences between the two distributions is to compare the

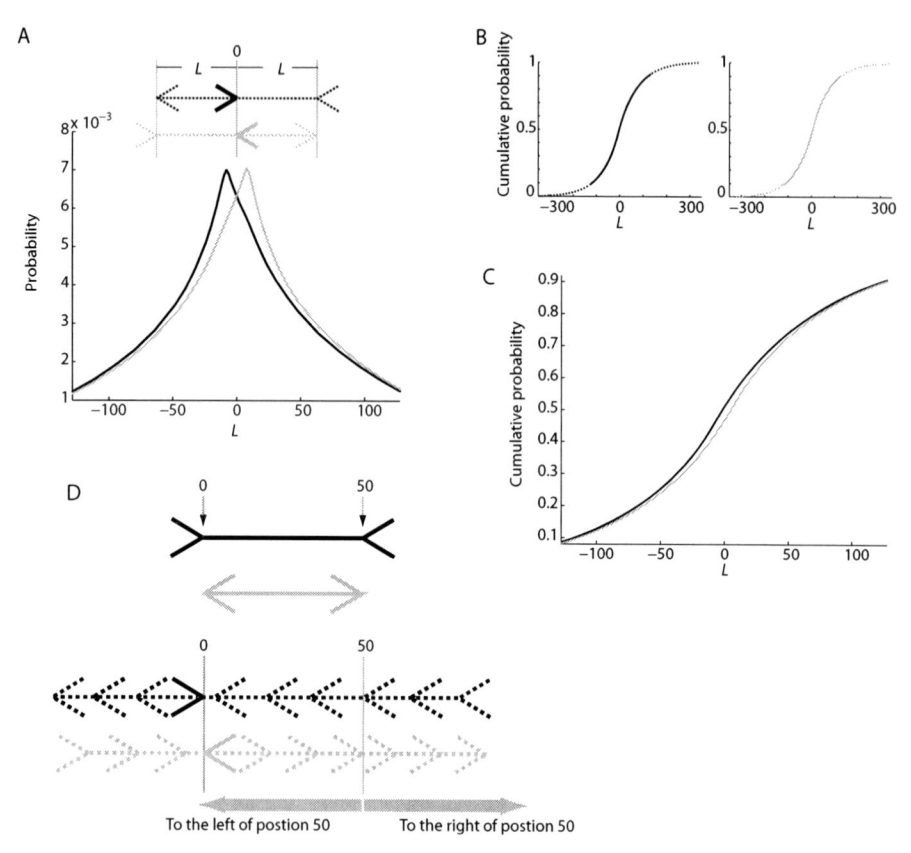

Figure 7.3 Statistical analysis of the physical sources of the standard Müller-Lyer stimulus, carried out on fully natural scenes. A) Probability distributions of the physical sources of Müller-Lyer figures with various shaft lengths (L, in pixels), given the presence of a conditional arrow adornment with its apex pointing either to the right (black) or to the left (gray). In the diagram above, the conditional adornments are indicated by solid lines, and the complementary adornments by dotted lines. B) The cumulative probability distributions derived from the probability distributions in (A) (the dotted portions of the curves are extrapolated). C) Superimposition of the two cumulative probability functions in (B). D) Examples of two shafts 50 pixels in length, one adorned with arrow tails and the other with arrowheads (upper panel). The left adornments are arbitrarily designated the conditional adornments, and are indicated by the solid elements at position 0 in the lower panel. Given each of these conditional adornments, the probability distributions shown in (A–C) indicate that the complementary adornment and shaft (dotted lines) occur at different positions with different probabilities. As a result, the summed probability of occurrence of all possible complementary adornments to the left of position 50 is greater when the fins of the conditional adornment extend to the left of position 0 (black) than when they extend to the right (gray). This statistical fact means that a complementary adornment at position 50, given a conditional adornment extending to the left of position 0, is further to the right in the empirical range of possible positions for complementary adornments than a complementary adornment at position 50 given a conditional adornment extending to the right. (After Howe and Purves, 2005b)

corresponding cumulative probability distributions (Figures 7.3B and C). The cumulative probability value for any given shaft length is the summed probability of occurrence of the physical sources of Müller-Lyer figures with shaft lengths less than or equal to that particular shaft length. (Since the shaft lengths considered here can have negative or positive values, negative shaft lengths are considered to be "less" than positive lengths in the cumulative distribution; for example, a shaft length of −30 is less than a shaft length of 10 in this context, even though the absolute value of −30 is greater than 10.) As is apparent in Figure 7.3C, for any specific shaft length, the cumulative probability derived from the probability distribution in black in Figure 7.3A is always greater than the cumulative probability corresponding to the probability distribution in gray. This statistical difference means that the summed probability of occurrence of the physical sources of Müller-Lyer figures whose complementary adornment is to the *left* of any specific position x, given the physical presence at position 0 of an arrow adornment whose apex points to the right, will always be greater than the same cumulative probability given the presence at position 0 of an arrow adornment whose apex points to the left.

PERCEPTUAL IMPLICATIONS

To understand the perceptual implications of the differences between the probability distributions in Figure 7.3, consider, as an example, two identical shafts 50 pixels in length, one adorned with arrow tails, and the other with arrowheads (Figure 7.3D). Take one of the adornments on each shaft, the one on the left for instance, as the conditional adornment, and the position of its apex as 0 (the same argument of course applies if the right adornment is selected). The complementary adornments are thus at position 50. Given the conditional adornment in the arrow-tails configuration in Figure 7.3D (black), the summed probability of occurrence of the physical sources of Müller-Lyer figures whose complementary adornment is to the left of position 50 is greater than the comparable cumulative probability given the conditional adornment in the arrowheads configuration (gray). Conversely, the summed probability of occurrence of the physical sources of complementary adornments located to the right of position 50, given the conditional adornment in the arrow-tails configuration, is less than the comparable probability given the conditional adornment in the arrowheads configuration.

These differences in the cumulative probabilities of the stimulus sources mean that the complementary adornment that actually occurred at position 50 in the arrow-tails configuration is further to the *right* in the empirical range of possible positions of complementary adornments than the complementary adornment that occurred at position 50 in the arrowheads configuration. These empirical differences thus predict that the complementary adornment in the arrow-tails configuration should appear further separated from the conditional adornment than the interval between the complementary adornment and the conditional adornment in the arrowheads configuration. Thus the shaft connecting the two adornments in the arrow-tails configuration in Figure 7.3D should be perceived as longer than the same line in the arrowheads configuration.

The same reasoning can be generalized to Müller-Lyer figures having any shaft length, meaning that a shaft adorned with arrow tails should always look longer than the same shaft adorned with arrowheads. This prediction is, of course, consistent with the percepts elicited by the standard Müller-Lyer stimulus.

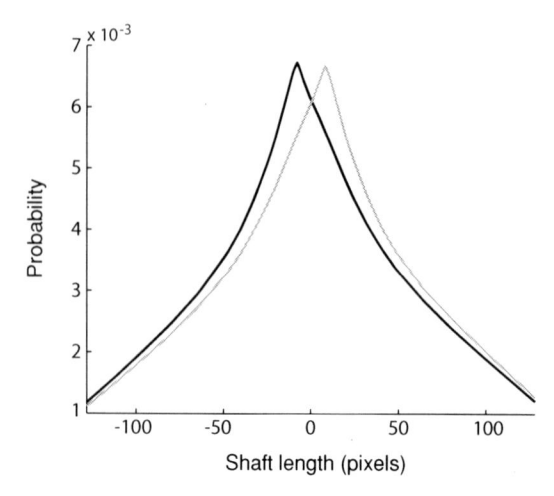

Figure 7.4 Probability distributions of the physical sources of the standard Müller-Lyer stimulus derived from scenes that include human constructions (cf. Figure 7.3A). (After Howe and Purves, 2005b)

STATISTICS DERIVED FROM DIFFERENT TYPES OF SCENES

The results presented so far were derived from fully natural scenes, presumably representing the dominant environment during the evolution of human visual perception. It was nevertheless of interest to carry out the same analyses on the scenes in the database that included human constructions, since the prevalence of rectilinear structures in man-made environments have sometimes been considered an important factor in the genesis of the Müller-Lyer effect (see below).

The probability distributions of the physical sources of the standard Müller-Lyer stimulus derived from this type of environment are shown in Figure 7.4. There is no obvious difference among the probability distributions obtained from this subset of the range data and the distributions derived from fully natural scenes (see Figure 7.3A).

VARIANTS OF THE MÜLLER-LYER STIMULUS

An obstacle for many previously proposed explanations of the Müller-Lyer illusion has been the similar perceptual effect elicited by the major Müller-Lyer variants, including shafts adorned with squares, standard arrow adornments without shafts and configurations comprising dots only (see Figure 7.1B–D). It was important, therefore, to carry out a similar statistical analysis using sampling templates in these configurations.

The probability distributions of the physical sources of these several variants are shown in Figure 7.5. In each case, the probability distributions derived in the context of the various adornments differ in the same general way as the probability distributions of sources determined for the standard Müller-Lyer stimulus illustrated in Figure 7.3. Thus the similar perceptual effect elicited by each of these variants can be rationalized in the same statistical framework as the standard stimulus.

The Müller-Lyer Illusion

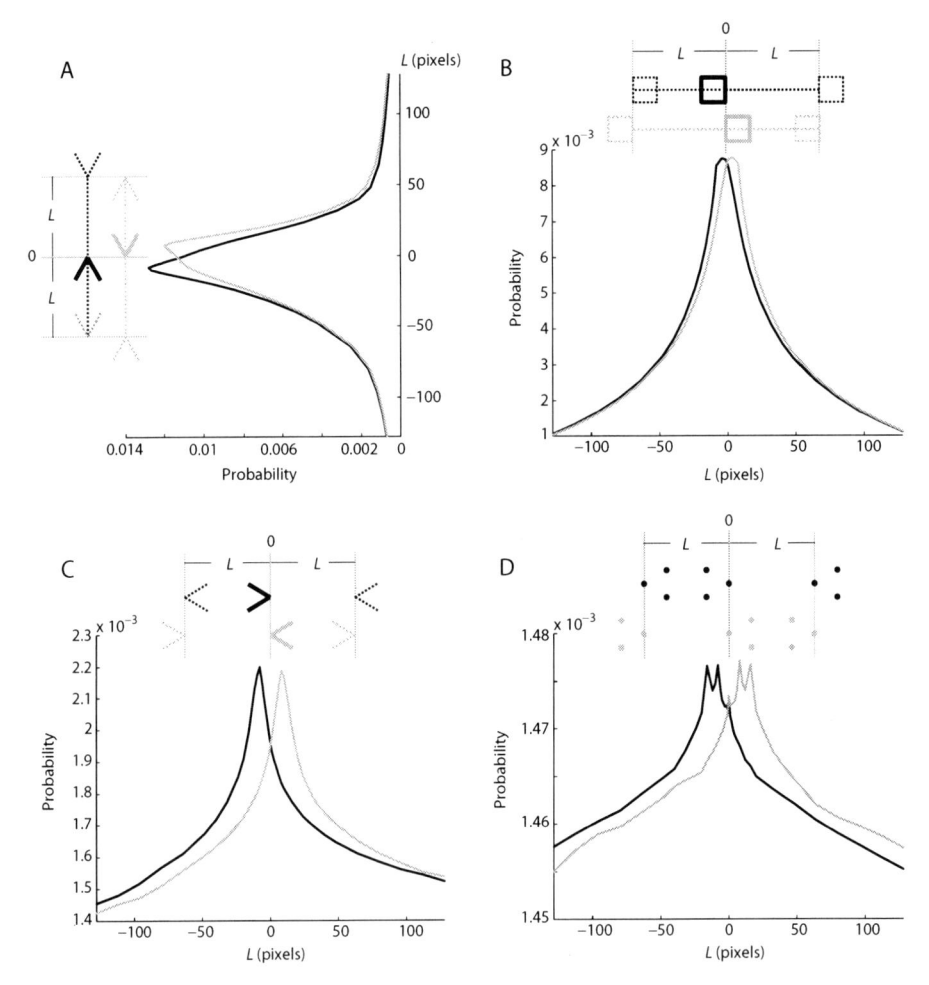

Figure 7.5 Probability distributions of the physical sources of the major Müller-Lyer variants. A) Standard Müller-Lyer stimulus oriented vertically. B) Müller-Lyer variant with square adornments. C) Variant with no shafts. D) Variant comprising only dots. The icons indicate the templates used to sample the database for each variant, in the same manner shown in Figure 7.3A. (After Howe and Purves, 2005b)

PHYSICAL BASIS FOR DIFFERENCES AMONG THE OBSERVED PROBABILITY DISTRIBUTIONS

The variations in the probability distributions of the physical sources of the Müller-Lyer stimulus according to the configuration of the adornments must arise from physical differences among these sources. What, then, is the nature of the differences?

Because straight lines in projected images, as emphasized earlier, typically arise from planar surfaces in 3-D space, the presence of the physical source of a conditional adornment comprising straight lines signifies the presence of a plane in the relevant region of the physical scene. Given this fact, the probability of occurrence of the physical

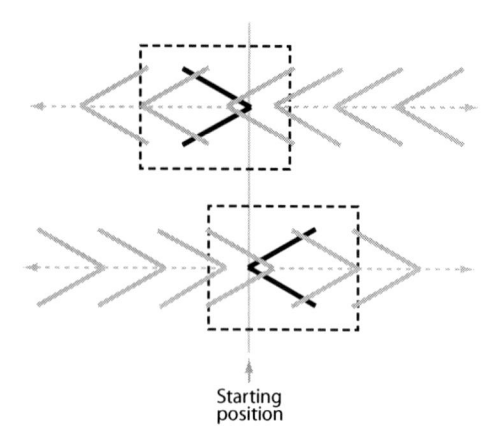

Figure 7.6 Physical basis for the differences in the probability distributions of the sources of Müller-Lyer stimuli. Given the presence of a conditional adornment (black) extending to the left of a starting position, the physical points corresponding to the complementary component (gray) would, on average, "fall off" the planar surface that contains the physical source of the conditional adornment sooner when moving away to the right than when moving away to the left. The opposite is true when the conditional adornment extends to the right of the starting position (the relevant planes are indicated by the dashed black lines).

source of *any* complementary component of the Müller-Lyer stimulus decreases as the interval between the two adornments increases. The reason is that as the complementary adornment becomes further separated from the conditional adornment, the physical points corresponding to the complementary component are less likely to be in the same plane as the physical source of the conditional adornment. Thus, in the presence of the physical source of a conditional adornment such as an arrow adornment whose fins extend to the left of a starting position, the physical points corresponding to the complementary component are less likely to be found in the plane of the conditional adornment when moving away to the right from that starting point than when moving away to the left (Figure 7.6, upper diagram). The opposite is of course true in the presence of a conditional adornment extending to the right of the starting position (Figure 7.6, lower diagram).

Thus the basis for the different probability distributions of the physical sources of the various Müller-Lyer figures considered here is the statistical difference in the likelihood of occurrence of the physical sources of the complementary components given the presence of the physical sources of different conditional adornments.

PREVIOUS EXPLANATIONS OF THE MÜLLER-LYER EFFECT

Two explanations of the Müller-Lyer illusion that have received significant attention over the years are the eye-movement theory (reviewed in Carr, 1935) and the assimilation theory (Pressey, 1967, 1970). The eye-movement theory claims that the misperception of the central shaft arises from the different extents of the eye movements needed to view a figure adorned with arrow tails compared to a figure with arrowheads. This older proposal has generally been dismissed because the illusion has been shown to

The Müller-Lyer Illusion

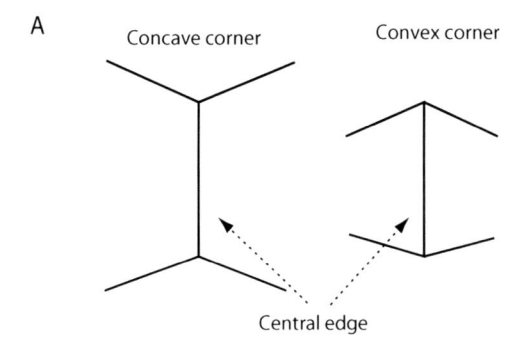

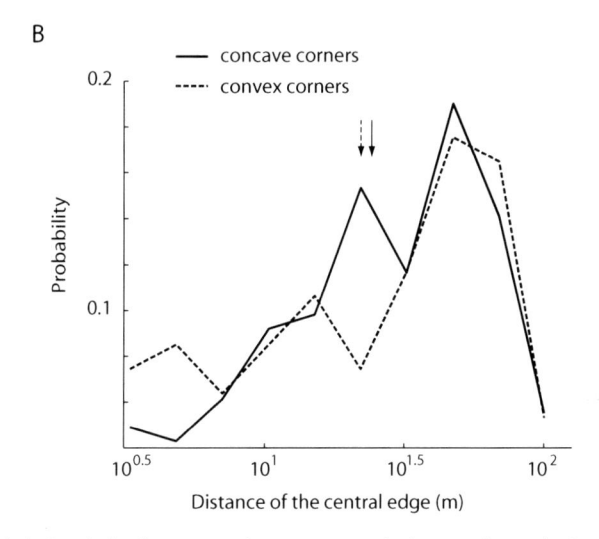

Figure 7.7 Statistical analysis of concave and convex corners in the range image database. A) Diagram of these two types of 3-D corners. B) Probability distributions of the distance from the image plane of the central edges of concave and convex corners in the database. Arrows indicate the means of the two distributions, which are not significantly different. (After Howe and Purves, 2005b)

persist in the absence of eye movements (Evans and Marsden, 1966; Bolles, 1969). The assimilation theory argues that the length of the central shaft is misperceived because the visual system cannot successfully isolate parts from wholes. In this scenario, the central shaft of the figure with arrow tails is seen as longer because the stimulus is, in its totality, longer. This explanation is contradicted, however, by the size contrast effects described in Chapter 5. Recall that a target embedded in a large surround (as in the Ebbinghaus stimulus, for example) appears *smaller* than the same target in a less extensive surround.

Of the previously suggested explanations of the Müller-Lyer illusion, the most cited is Richard Gregory's proposal that the stimulus with arrow tails signifies a concave corner in the 3-D world, whereas the figure with arrowheads signifies a convex corner, the central shaft corresponding to the central edge of the two types of corners (Gregory, 1963, 1966) (Figure 7.7A). Since the central edge of concave corners is taken to be further from an observer than the central edge of convex corners, the central shaft of the Müller-Lyer figure with arrow tails would, in this interpretation, appear longer as a "compensation" for the different average distances of the 3-D corners they represent.

This explanation of the Müller-Lyer effect has been rejected by some investigators because it does not explain the effects elicited by the variants illustrated in Figures 7.1B–D (Rock, 1995), and because the effect persists even when a context rich in depth information contravenes any difference in the distances of the central shafts (see, for example, Figure 1.3A). It was nonetheless of interest to examine the merits of this influential idea directly. Accordingly, all the 3-D corners in the range image database were identified by visual inspection and the distance to the central edges of the two types of corners measured. Although only a small number of samples were obtained in this way (~200 samples for each type of corner), there was no significant difference between the probability distributions of the distances from the image plane to the central edges of concave and convex corners (Figure 7.7B). Thus, although Gregory's intuition about the empirical significance of the Müller-Lyer stimulus points in the right general direction (i.e., an explanation based on past experience with the sources of such stimuli), convex and concave corners as such contribute little or nothing to the genesis of the Müller-Lyer effect.

SUMMARY

The results summarized here further support the idea that visual percepts are determined by the statistical relationship between retinal images and their possible real-world sources. The otherwise puzzling perceptual effects of the standard Müller-Lyer stimulus and the several variants that have been especially difficult to explain are evidently another signature of this wholly probabilistic strategy of vision.

Chapter 8

The Poggendorff Illusion

Another complex geometrical illusion that has been the subject of much study and debate is the Poggendorff effect, which was first described in the middle of the 19th century (see Figures 1.2E and 1.3B). Johann Poggendorff, a chemistry professor at the University of Berlin, pointed out that when the continuity of an obliquely oriented line is interrupted, the position of the line segments on either side of the interruption appear to be shifted vertically (or horizontally, if the interruption is oriented horizontally) (the history of Poggendorff's creation is reviewed in Robinson, 1998, p.76 ff).

THE POGGENDORFF EFFECT

This intriguing phenomenon is typically presented in the form shown in Figure 8.1A or 8.1B. In the format in Figure 8.1A, the right oblique line segment appears to be shifted downward relative to the left segment, even though they are actually collinear. Figure 8.1B is the mirror image of Figure 8.1A; in this case the right line segment appears to be shifted upward relative to the left segment. A similar effect is elicited when the oblique line is interrupted by a space delineated by two parallel horizontal lines instead of two vertical lines (Figure 8.1C). In this presentation, the right segment appears to be shifted to the right from the position of collinearity with the left segment. The magnitude of the apparent shift of the oblique line segments in these several configurations also varies as a function of both the orientation of the interrupted line and the width of the interruption. In the standard presentation of the Poggendorff stimulus, such as shown in Figures 8.1A and B, the effect increases as the orientation of the interrupted line becomes increasingly vertical (Figure 8.1D)(Weintraub and Krantz, 1971; Day and Dickinson, 1976); the effect also increases as the width of the interruption is increased (Figure 8.1E) (Weintraub and Krantz, 1971).

A particularly puzzling aspect of the phenomenon is that the effect is largely abolished when only the acute components of the standard Poggendorff stimulus are shown. However, the effect remains if only the obtuse components are present (Day, 1976) (Figure 8.1F). Finally, when the overall orientation of the stimulus is rotated

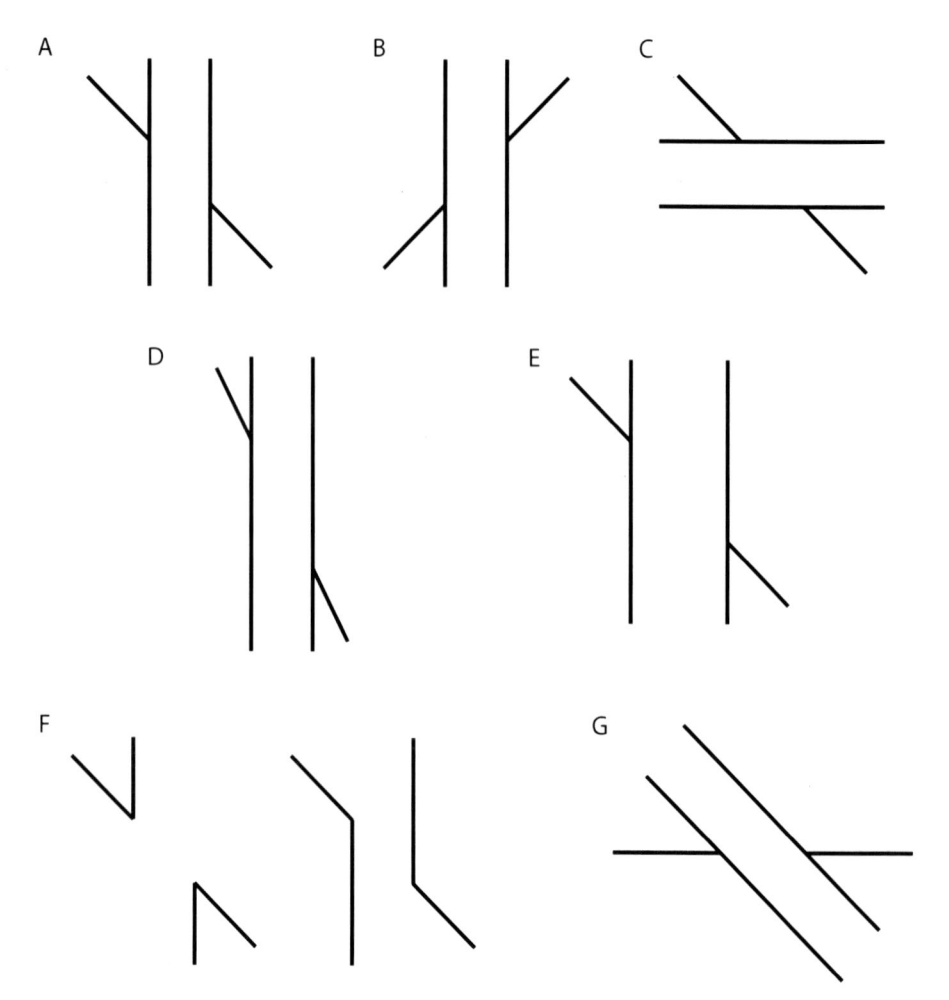

Figure 8.1 The Poggendorff illusion and its behavior. A) When an obliquely oriented straight line is interrupted by a vertical "occluder", the line segment on the right appears to be shifted downward with respect to the line segment on the left. B) A similar effect occurs when the orientation of the interrupted line is reversed. In this case, the collinear extension on the right appears to be shifted upward. C) When an oblique line is interrupted by a space delineated with parallel horizontal lines, the oblique line segments appear to be shifted horizontally with respect to each other. D) The magnitude of the effect increases when the interrupted line becomes closer to vertical. E) The magnitude also increases as the width of the interruption increases. F) The illusion is largely abolished when only the acute components of the stimulus are presented, but maintained when only the obtuse components are shown. G) The illusion is diminished when the standard configuration is rotated so that the interrupted line is horizontal. (After Howe et al., 2005)

such that the interrupted line is horizontal (Figure 8.1G), the effect is reduced but not completely abolished (Leibowitz and Toffey, 1966; Weintraub and Krantz, 1971; Day and Dickinson, 1976).

Although many theories have been proposed over the past 150 years to account for the Poggendorff effect (see below), there is, as in the case of the Müller-Lyer stimulus and its variants, no explanation that rationalizes the full range of behavior illustrated in Figure 8.1.

STATISTICAL ANALYSIS OF THE PHYSICAL SOURCES OF POGGENDORFF STIMULI

When the continuity of a physical object is interrupted by occlusion, the location of its re-appearance on the other side of the interruption is uncertain: there are many ways the object could have traversed the interval, and the interruption does not provide any information about which of these possible ways underlies the stimulus. The oblique line segments in the standard Poggendorff presentation in Figure 8.1 could arise from a single object in 3-D space with an infinite variety of possible configurations of the part of the object behind the implied occluder, or from two different objects altogether. The Poggendorff stimulus and its variants thus reflect the fundamental uncertainty of how the physical sources of projected line segments are actually continued across a spatial interval. From this perspective, the percepts elicited by these stimuli should reflect the past experience of human observers with the physical sources of straight-line segments that are in the same orientation but separated by a spatial interval. In other words, given a line segment on one side of an implied occluder, the perceived position of another line segment in the same orientation should be determined by the relative probability of occurrence of the physical sources of line segments that have projected in that orientation across the occluder.

Figure 8.2A shows examples of the geometrical templates applied to the range images to sample the physical sources of the four lines comprising the standard Poggendorff stimulus. The determination of the physical sources of the stimulus involved two steps. First, we identified a region of a scene that contained a physical source of one of the two oblique line segments (the left segment, for example), as well as the sources of the two vertical lines. Then, in the same region of the scene, we determined the frequency of occurrence of the physical sources of possible right line segments, i.e., line segments that had the same projected orientation as the left oblique segment, and that were located just to the right of the right vertical line.

A template for sampling the left oblique line segment and the two vertical lines is shown in Figure 8.2B. As described in previous chapters, the points in the image underlying each straight line in the template were examined to determine if the corresponding physical points formed a straight line in 3-D space. If so, the set of physical points identified by the template was accepted as a valid physical source of these three components of the stimulus. After identifying a region of a scene that contained the sources of the first three lines, the template for sampling the right oblique line segment was applied just to right of the right vertical line (see Figure 8.2A and B). This template was moved vertically in sequential applications to determine the occurrence of all the possible physical sources of the right oblique segment at different vertical locations relative to an extension of the left oblique segment. The location of the right segment is described in terms of the distance (measured by the visual angle subtended) from the point of intersection of the right oblique segment with the right vertical line to the

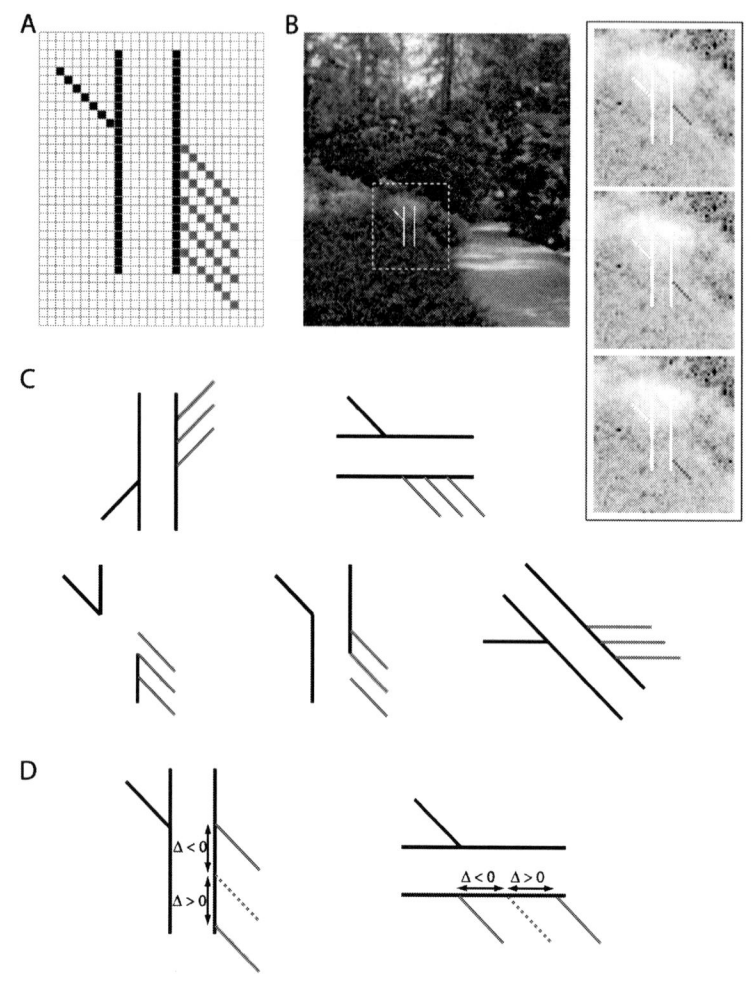

Figure 8.2 Determining the possible physical sources of the right line segment in a Poggendorff stimulus, given the other three lines. A) Examples of the templates for sampling the lines comprising the standard Poggendorff stimulus (see Figure 8.1A) in the range image database. The pixels in an image patch are represented diagrammatically by the grid squares; black pixels indicate the template for the left oblique line segment and the two vertical lines; the red pixels indicate a series of templates for sampling the right line segment at various vertical locations relative to the left oblique segment. As in previous analyses, the points underlying a template were accepted as a valid sample only if they formed a straight line in 3-D space. B) The solid white lines indicate a valid sample of the left oblique line segment and the two vertical lines. Blowup of the boxed portion of the scene shows the subsequent application of a series of templates (red) used to test for the presence of right oblique segments at different vertical positions. C) Using the different sampling templates shown here, the possible sources of all the various configurations of the Poggendorff stimulus illustrated in Figure 8.1 could be determined. In these further examples, the black lines are equivalent to the black template in (A), and the red lines equivalent to the red templates. D) Definition of the difference in the physical locations of the two segments of the interrupted line. The symbol Δ signifies the location of the right line segment relative to the location at which the left line segment, if extended (dotted red line), would intersect the right vertical line (or the lower horizontal line); Δ is arbitrarily designated negative if the right segment is located above (or to the left of) this point of intersection, and positive if otherwise. (After Howe et al., 2005)

point at which the left oblique segment, if extended, would intersect the right vertical line (see Δ in Figure 8.2D).

The physical sources of the variants of the Poggendorff stimulus illustrated in Figure 8.1 were similarly determined using appropriately configured sampling templates, as shown in Figure 8.2C. For each of the configurations tested, the number of occurrences of the physical sources of the right line segments at different locations relative to the left segment was tallied, and this information used to produce the corresponding probability distribution of the sources of the right segments. We could then ask if these distributions predicted the variety of perceptual phenomena elicited by the standard Poggendorff stimulus and its variants.

STATISTICAL EXPLANATION OF THE MAIN EFFECT ELICITED BY THE POGGENDORFF ILLUSION

As shown in Figure 8.3, the probability distribution derived in this way from the fully natural scenes in the database can indeed account for the main feature of the Poggendorff illusion. When the left oblique segment is oriented downward and to the right, as in Figure 8.1A, the probability distribution of the possible sources of the right segment is biased toward locations that correspond to negative values of Δ (Figure 8.3A). This result means that a majority of the physical sources that give rise to projections of right oblique line segments in the same orientation as the left segment will, in past human experience, have projected *above* the point at which an extension of the left oblique line intersects the right vertical line.

The frequency of the occurrences of physical sources of right line segments at different vertical locations would, according to the statistical framework of vision proposed here, have influenced how observers perceive the relative positions of the left and the right line segments in the Poggendorff stimulus. To understand the nature of this empirical influence, consider a hypothetical probability distribution of the physical sources of right line segments in which the sources at all vertical locations are equally likely. In this case, the perceived vertical location of any right line segment predicted by these probabilities (i.e., the empirical rank of the position of the line) would be the same as its geometrical location. Thus, a right line segment that was actually collinear with the left segment would be seen as such.

The real-world probability distribution of the sources of the right line segment shown in Figure 8.3A, however, is not uniform. The statistical fact that more physical sources of right line segments have projected above the point of collinearity with the left segment than below it in the past experience causes the position of a right line segment that is geometrically collinear with the left segment to be shifted *downward* in the empirical space determined by these statistics. Accordingly, the right oblique line segment in the standard Poggendorff stimulus in Figure 8.1A should appear shifted downward and thus somewhat below the point where an extension of the left segment would intersect the right vertical line. This is indeed what people see.

When, however, the orientation of the interrupted line in the stimulus is upward from the lower left, as in Figure 8.1B, the probability distribution of the sources of the right line segment is biased toward locations *below* the point at which a continuation

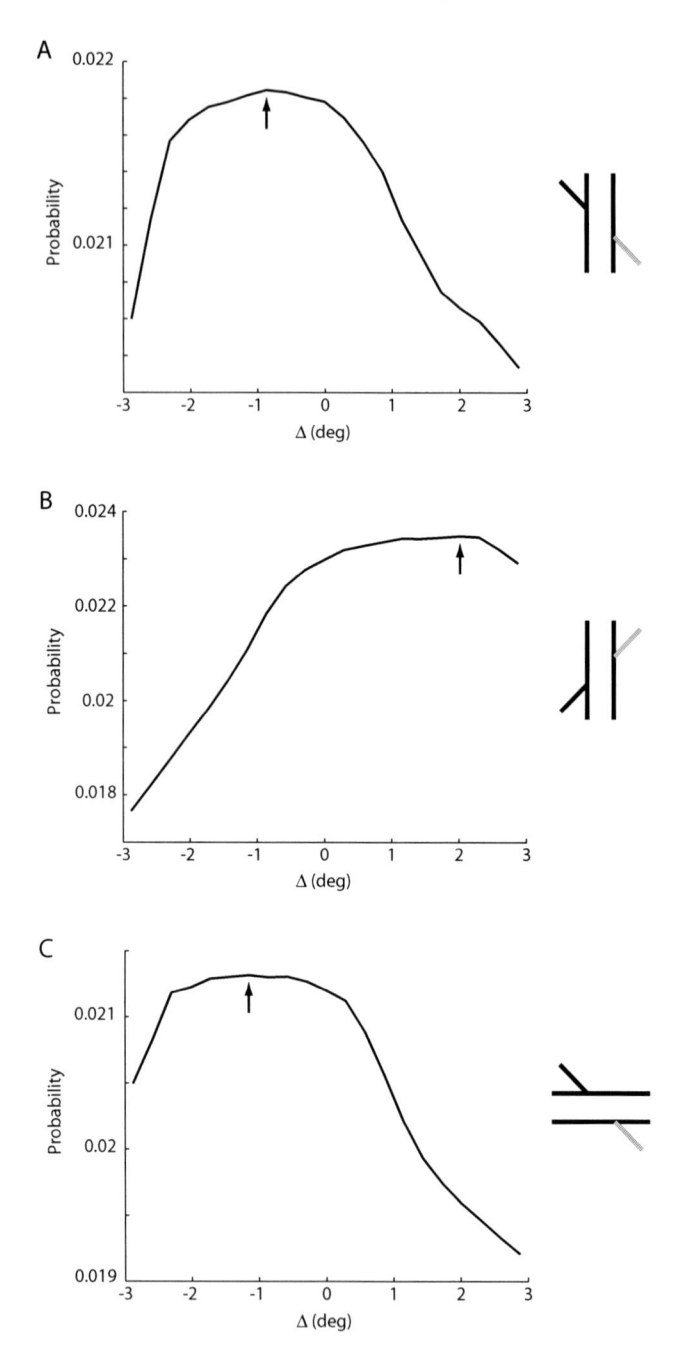

Figure 8.3 Probability distributions of the physical sources of the right line segment in the standard Poggendorff stimuli. A) The probabilities of occurrence of the physical sources of the right oblique segment for the standard configuration of the Poggendorff stimulus in Figure 8.1A, presented as a function of the location of the right segment relative to the left oblique segment (Δ, measured in terms of visual angle). Arrow indicates the mode of the distribution. B, C) Probability distributions of the physical sources of the right line segment for the stimulus configurations in Figures 8.1B and 8.1C, respectively. (After Howe et al., 2005)

The Poggendorff Illusion

of the left line intersects the right vertical line (i.e., toward values of $\Delta > 0$) (Figure 8.3B). By the same reasoning, these statistics predict that a right oblique segment that is actually collinear with the left segment will in this case be seen as located *above* the point where an extension of the left segment would intersect the right vertical line, as it is.

The same sort of explanation also applies to the Poggendorff stimulus in which an oblique line is interrupted by two horizontal lines. When the left oblique segment is oriented downward and to the right, as in Figure 8.1C, the probability distribution of the sources of the right line segment is biased toward locations to the *left* of the point where $\Delta = 0$ (Figure 8.3C). This distribution thus predicts that the apparent location of the right segment should be shifted to the *right* from the point at which a continuation of the left line actually intersects the lower horizontal line. This statistical result is again consistent with what people see.

ADDITIONAL FEATURES OF THE ILLUSION EXPLAINED IN THESE TERMS

The effects of varying the intersecting angle of the interrupted line (see Figure 8.1D) and of varying the width of the interruption (see Figure 8.1E) on the corresponding probability distributions of the physical sources of the right line segment are shown in Figures 8.4A and 8.4B, respectively. Changing either the orientation of the interrupted line or the width of the interruption systematically shifts the probability distribution in a manner consistent with the changing magnitude of the perceptual effects in these presentations. As the orientation of the interrupted line becomes closer to the orientation of the lines defining the interruption (vertical in the example in Figure 8.4A), or as the

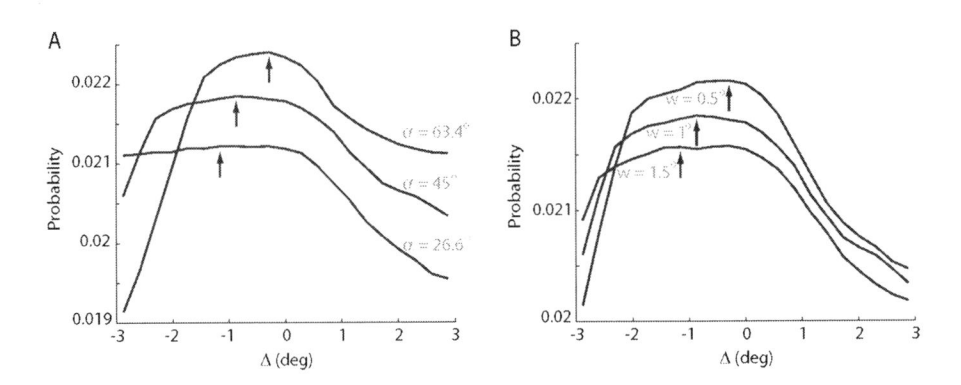

Figure 8.4 The probability distribution of the physical sources of the right line segment changes progressively as the orientation of the interrupted line or the width of the interruption is altered. A) The probability distributions of the sources of the right line segment when the angle of intersection (α) between the oblique line segments and the vertical lines is 63.4°, 45° or 26.6° (in each case the width of the interruption was 1° of visual angle). B) The probability distributions of the sources of the right line segment when the width of the interruption (w) is 0.5°, 1° or 1.5° of visual angle (the angle of intersection was always 45° in this example). Arrows indicate the modes of the distributions. (After Howe et al., 2005)

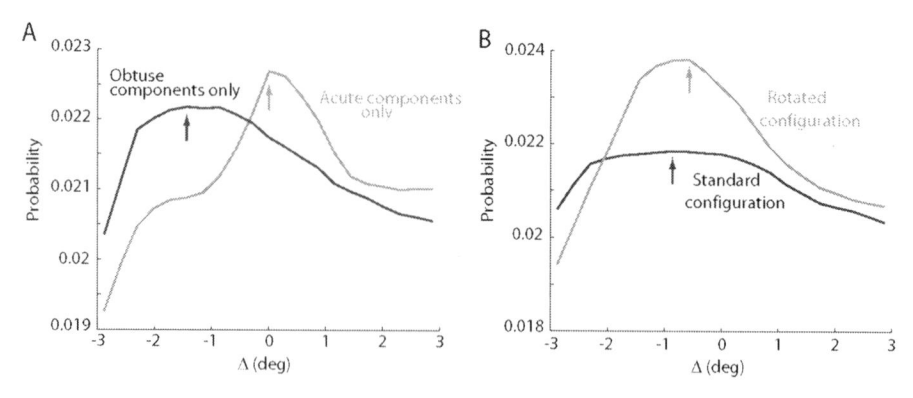

Figure 8.5 Probability distributions of the physical sources of the right line segment when the standard Poggendorff stimulus is decomposed or rotated. A) The probability distribution obtained when only the acute components of the stimulus are sampled, compared to the distribution when only the obtuse components are considered. In both cases the orientation of the interrupted line was 45° and the width of the interruption 1°. B) The probability distribution of the sources of the right line segment when the entire stimulus is rotated such that the interrupted line is horizontal (see Figure 8.1G). Arrows indicate the modes of the distributions. (After Howe et al., 2005)

interruption becomes wider, the mode of the distribution shifts progressively away from the point where $\Delta = 0$, in accord with the fact that the perceived shift in the apparent location of the right line segment increases as the angle of the intersection becomes smaller, or as the width of the interruption increases (see Weintraub and Krantz, 1971).

Another puzzle apparent in the perceptual effects illustrated in Figure 8.1 is why, in these terms, the Poggendorff effect is greatly diminished when only the acute angles in the standard stimulus are presented, but little affected when the presentation is restricted to the obtuse components (see Figure 8.1F). As shown in Figure 8.5A, when only the acute elements are used for the templates sampling the range images (see Figure 8.2C), the mode of the probability distribution (arrow) is very near the point where Δ equals zero. Accordingly, the Poggendorff effect should be largely abolished, as it is. Conversely, when only the obtuse components are used, the probability distribution is similar to that of the standard Poggendorff stimulus; thus the illusion would be expected to retain its full magnitude, as it does.

Finally, we determined the probability distribution of the sources of the right line segment when the overall orientation of the Poggendorff stimulus is rotated so that the interrupted line is horizontal, as in Figure 8.1G. Compared to the probability distribution of the sources of the standard presentation shown in Figure 8.3A, the distribution when the stimulus is rotated 90° has a mode closer to $\Delta = 0$, as shown in Figure 8.5B. Accordingly, the Poggendorff effect should be reduced when the presentation of the interrupted line is horizontal, as is again the case.

In sum, the probability distributions of the physical sources of line segments in both the standard Poggendorff stimulus and its variants in natural scenes account not only for the main Poggendorff effect, but also the percepts elicited by the stimulus in a variety of other presentations that are otherwise difficult to explain.

The Poggendorff Illusion

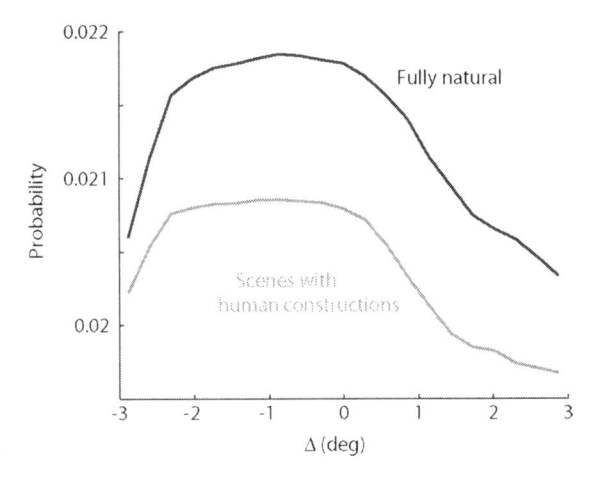

Figure 8.6 The probability distributions of the sources of the right line segment for the standard Poggendorff stimulus derived from an analysis of fully natural scenes (see Figure 8.3A) compared to the distribution derived from the scenes that contained human constructions. (After Howe et al., 2005)

STATISTICS DERIVED FROM DIFFERENT TYPES OF SCENES

Theories purporting to explain one or more aspects of the Poggendorff or other complex geometrical illusions (see Chapter 7) have often been predicated on intuitions about the "carpentered" world of human artifacts (e.g., Gregory, 1997; Gillam, 1998; Changizi and Widders, 2002). It was thus of interest to ask how the probability distributions derived from fully natural scenes (presumably representative of the human visual environment during evolution) compare to distributions derived from environments in which human construction plays a large part. As in our consideration of the Müller-Lyer illusion, we computed the relevant probability distributions from the set of scenes in the database that included some or predominantly man-made objects.

As is apparent in Figure 8.6, the pertinent probability distributions obtained from the two types of scenes are generally similar. It thus seems safe to conclude that the perceptual effects apparent in the Poggendorff illusion are—like the Müller-Lyer effect—not specifically dependent on interactions with the rectilinear constructions associated with human culture.

PHYSICAL BASIS FOR THE STATISTICAL BIASES OBSERVED

Two questions so far deferred in considering the statistics derived from the analysis of range images are why the observed biases in the occurrence of the physical sources of the right line segment exist, and why the magnitude of these biases differs for the different configurations of the Poggendorff stimulus considered here.

The answer to these questions again lies in the geometry of planar surfaces, which, as discussed earlier, are the typical sources of straight-line projections on the retina.

94 Perceiving Geometry: Geometrical Illusions Explained by Natural Scene Statistics

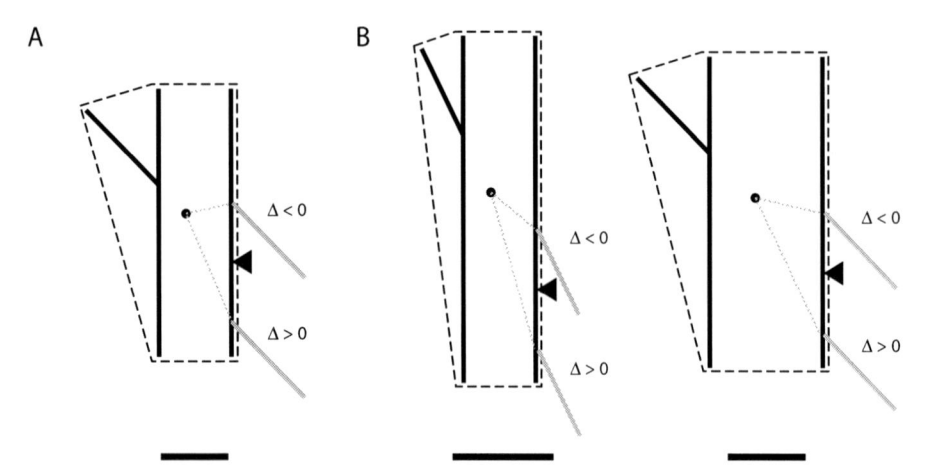

Figure 8.7 Physical basis for the bias in the probability distributions of the sources of Poggendorff stimuli. A) In the case of the standard stimulus, a right line segment above the point where $\Delta = 0$ (arrow head) is closer to the center of the plane containing the physical source of the left segment and the vertical lines than is an otherwise comparable right line segment below the point where $\Delta = 0$. The distances of the right line segments from the center of the plane (black dot within the dashed line that defines the plane) are indicated by the gray dotted lines; the difference in the lengths of the two gray dotted lines is indicated by the black bar below the diagram. As a result of this physical bias, the probability distribution of the sources of the right line segment is biased toward vertical locations above the point where $\Delta = 0$. B) As the orientation of the interrupted line becomes closer to vertical (left panel), or as the width of the interruption increases (right panel), the same amount of shift in the vertical position of the right line segment means a larger difference in the distances of the right line segments from the center of the plane. The difference in the lengths of the gray dotted lines is again indicated by the black bars below. The larger the difference in the distances of the right line segments from the center of the plane, the more pronounced the bias in the relevant probability distribution, and thus the more pronounced the perceptual effect. (After Howe et al., 2005)

Consider, for instance, the physical sources of the standard Poggendorff stimulus in Figure 8.1A. The presence of a physical source of the left oblique line segment and the vertical lines would usually signify the presence of a planar surface in the corresponding region of the scene. Since a right oblique line segment above the point at which Δ equals 0 is, on average, closer to the center of the plane than a line segment below this point (Figure 8.7A), the set of physical points corresponding to a right line segment is statistically more likely to be in this plane when it is above the point where $\Delta = 0$ than when it is below this point. Thus, the likelihood of occurrence of the physical sources of the right line segment is greater at positions where $\Delta < 0$ (i.e., above the point of $\Delta = 0$) than at positions where $\Delta > 0$. This is why the probability distribution in Figure 8.3A is biased towards $\Delta < 0$. The same reasoning can explain the biases seen in the distributions in Figure 8.3B and C.

When the orientation of the interrupted line in the standard Poggendorff stimulus is closer to vertical, or when the width of the interruption increases, the same shift in the vertical position of the right line segment means a larger *difference* in the distances of right line segments from the center of the plane containing the physical source of the left segment and the vertical lines (Figure 8.7B). There is, accordingly, an increased bias in the probability distribution of the sources of the right line segment as a function of the orientation of the interrupted line, and as a function of the width of the interruption (see Figure 8.4).

The reason for the greater bias in the presence of only the obtuse components of the stimulus compared to only the acute components (see Figure 8.5A) is also straightforward in these terms. In the stimulus comprising the acute components (see Figure 8.1F), the absence of right vertical line above the point where $\Delta = 0$ means that the planar surface giving rise to the left segment and the vertical lines is less likely to extend above this point compared to the sources of the standard stimulus. Thus, the likelihood of occurrence of the sources of the right oblique segment is no longer biased toward positions where $\Delta < 0$. In contrast, the right vertical line *is* present above the point of $\Delta = 0$ in the stimulus composed of the obtuse components; the bias toward $\Delta < 0$ in the corresponding distribution of the sources of the right oblique segment is therefore maintained.

Finally, the bias evident in the probability distribution of the sources of the right line segment is reduced when the standard Poggendorff stimulus is rotated such that the interrupted line is horizontal (see Figure 8.5B). The reason for the reduction in this case lies in the higher overall frequency of occurrence of the sources of horizontal lines in the physical world (see Chapter 3). (The reduction cannot be rationalized in terms of the relationship between the right line segment and the plane containing the source of the left segment and the vertical lines because this relationship is unchanged from that in the standard stimulus, despite the rotation.) The greater probability of occurrence of sources of horizontal lines compared to the sources of oblique lines makes the probability distribution of the right horizontal line segment in the rotated stimulus less susceptible to the bias caused by the presence of the other stimulus elements, much as a higher baseline makes a signal detector less sensitive to the same input.

OTHER EXPLANATIONS OF THE POGGENDORFF EFFECT

The Poggendorff effect has often been considered a manifestation of the misperception of the angles in the stimulus (Blakemore et al., 1970; Burns and Pritchard, 1971). For instance, an overestimation of the acute angles in the standard stimulus in Figure 8.1A and an underestimation of the obtuse angles would presumably bias the apparent orientation of the line segments on either side of the interruption toward horizontal, thus causing them to appear non-collinear. This explanation, however, has been disputed on the grounds that it cannot rationalize the paradoxical destruction and preservation of the effects elicited respectively by the presentation of only the acute or the obtuse components of the stimulus (see Figure 8.1F). Nor does this explanation account for the additional features of the Poggendorf effect apparent in Figures 8.1D, E and G.

Another type of explanation is based on how geometrical information in 2-D projections might be inappropriately "interpreted" by observers. For example, the "depth processing" theory (see Chapter 1) suggests that the oblique lines in the Poggendorff stimulus are interpreted as lines extending in depth, and are therefore perceived to be non-collinear (Gillam, 1971; see also Gregory, 1963, 1997). This sort of explanation is also limited in that it considers only some aspects of the projective geometry of the stimulus and its possible sources, and thus has difficulty explaining the full range of the behavior of the Poggendorff stimulus illustrated in Figure 8.1.

SUMMARY

One of the most controversial of the many discrepancies between perceived spatial relationships and the physical structure of a visual stimulus is the Poggendorff illusion, in which an obliquely oriented line that is interrupted no longer appears collinear. Despite its relative complexity, this illusion and its altered behavior in a variety of different presentations can all be rationalized by the statistical relationship between the stimulus and its possible sources in typical visual environments, in the same statistical framework as the analyses summarized in the preceding chapters.

Chapter 9

Implications

The previous chapters provide much evidence that the percepts observers see in response to geometrical stimuli are determined by the probability distributions of the possible sources of the stimulus in question. A corollary is that visual space and all its nuances (i.e., all the geometrical aspects of the world that we perceive by means of vision) are equally a result of this fundamentally probabilistic strategy of vision. As outlined in Chapter 1, the biological rationale for this way of perceiving the world is straightforward in principle: because the sources of geometrical stimuli are inevitably ambiguous, and because behavior must contend with the sources of stimuli rather than the stimuli per se, this strategy is really the only plausible way human beings or other visual animals can cope with the inverse optics problem. This conclusion, however, does not address how the nervous system implements this solution, i.e., what implications this strategy has for understanding the properties of visual neurons, the circuits they form and the anatomical features that their collective organization generates. The purpose of this last chapter is to examine some of the major views that have been held previously and what the present evidence adds to these perspectives.

THE ORIGINAL CONCEPT OF NEURONAL RECEPTIVE FIELDS

Ever since David Hubel and Torsten Wiesel's groundbreaking work that began more than 40 years ago (Hubel and Wiesel, 1959, 1962), many vision scientists have assumed that the function of the visual brain, at least with respect to relatively simple aspects of vision such as responding to the basic geometry of scenes, is to extract and encode the relevant features of physical objects (e.g., their linear dimensions, angular subtenses, orientation of contrast boundaries and distances).

The general idea has been based, by and large, on the seminal series of electrophysiological observations that Hubel and Wiesel initiated in the late 1950s and that have been carried forward by them and many others. Neurons at the input stages of the visual pathway (retina, lateral geniculate nucleus [LGN] of the thalamus and the primary visual cortex [V1] in the occipital lobe) were found to be selective for stimuli presented at a particular location in the visual field of the observer (a cat or monkey in

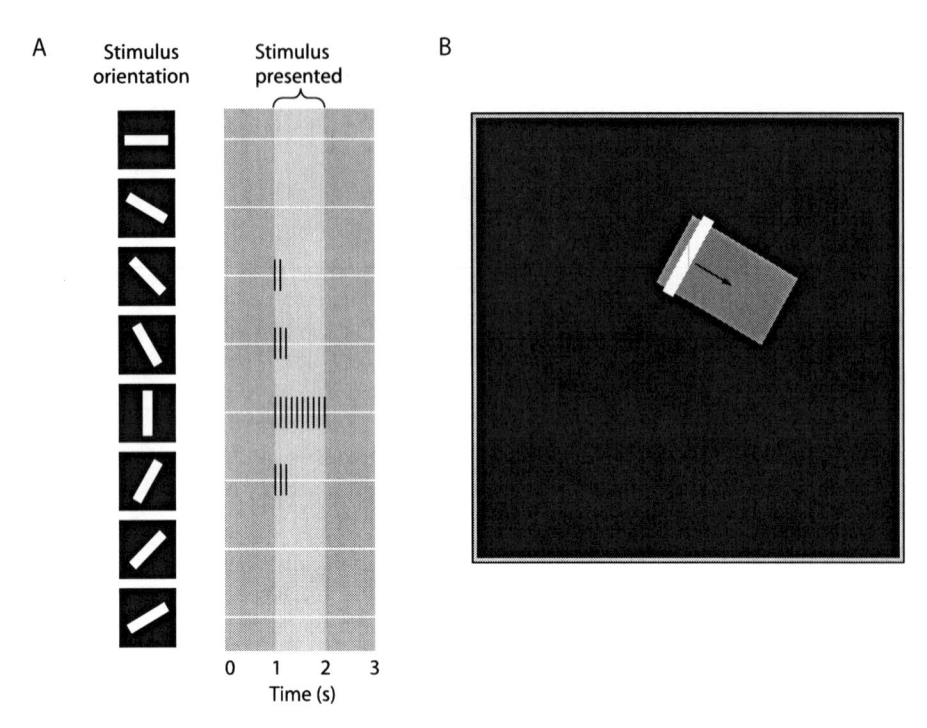

Figure 9.1 Schematic representation of a visual neuron's receptive field properties determined by electrical recording from a single cell in the primary visual cortex of a cat or monkey. A) The firing rate of the neuron illustrated here as an example varies as a function of the orientation of the line (a bar of light moved across a darkened screen). B) The area on the screen (indicated in gray) and the qualities of the stimulus (e.g., the orientation, direction and speed of the moving bar) required to significantly change the firing rate of the neuron from baseline define that neuron's "classical" receptive field properties. (After Purves and Lotto, 2003)

these experiments), a fact that defines the overall location of each cell's *receptive field*. At each of these levels of the visual system, receptive fields are arranged topographically such that adjacent receptive fields correspond to adjacent positions on the retina and thus in the visual space.

Although neurons in the retina and lateral geniculate nucleus are activated by spots of light in the appropriate region of the visual field, neurons in V1 are more vigorously activated by luminance edges at particular orientations (Figure 9.1). This finding implies that the response properties of neurons at this initial processing level of the cortex are further defined by a set of specific stimulus characteristics that change the neuron's activity from its baseline rate of firing. These characteristics, together with the topographical location, are referred to as the *receptive field properties* of a neuron.

The receptive field properties of visual neurons at many levels of the central nervous system have now been studied by numerous investigators in a variety of species over several decades. In addition to being selectively responsive to edges of different orientations, neurons in V1 and the extrastriate visual cortex of experimental animals such as cats and monkeys respond selectively to the direction of movement of the stimulus (Hubel and Wiesel, 1968; Movshon, 1975; De Valois et al., 1982), the speed of stimulus movement (Maunsell and Van Essen, 1983; Movshon et al., 1986; Van Essen and Gallant, 1994; Recanzone et al., 1997), its spectral characteristics (Zeki, 1983a, b; Thorell et al., 1984; Tootell et al., 1988) and the binocular disparity of the stimulus (Barlow et al., 1967; Hubel and Wiesel, 1977; Poggio et al., 1988). Moreover, some neurons in visual cortical areas in the temporal lobe respond selectively to higher-order stimulus characteristics such as shape and texture (Baylis et al., 1987; Tanaka, 1993), and even to complex patterns associated with specific objects or faces (e.g., Desimone and Gross, 1979; Bruce et al., 1981; Rolls 1984, 1994).

These studies have all been based on, and in turn have reinforced, the idea that the primary function of visual neurons is to detect various "features" in retinal images, and that the characteristic "tuning properties" of neuronal receptive fields exemplify this goal. With respect to perception, the supposition, whether implicit or explicit, is that neuronal receptive field properties are translated into visual percepts, and that what we ultimately see is determined by the receptive field properties of all the neurons activated (or inactivated) by a given stimulus.

THE GENESIS OF RECEPTIVE FIELD PROPERTIES

Given this evidence, vision scientists have naturally asked how visual processing circuitry at different levels of the system gives rise to the relevant receptive field properties. The idea underlying this effort has generally been that perception must be rationalized in terms of feature extraction and representation, and that understanding how the selective properties of receptive fields are constructed will eventually elucidate how this is accomplished at the level of individual neurons or small groups of neurons.

A good example of this approach is again found in the studies carried out by Hubel and Wiesel (1962, 1974, 1977). As is now described in virtually all neuroscience textbooks, Hubel and Wiesel showed that by an alignment of their receptive fields, inputs to V1 from neurons in the LGN (which are not themselves selective for orientation) could combine to account for the orientation selectivity of neurons in the primary visual cortex. This generation of new properties at a higher level of the visual pathway implies that the output of orientation-selective cells in V1 (sometimes called "edge detectors") can be combined to generate a representation of luminance contours in the retinal image at still higher levels of the system. Pursuing this idea, such contour-detecting neurons could give rise to representations of object shapes at even higher cortical levels. The combined activity of neurons at the apex of this series of ever more complex representations would presumably correspond to what we actually see.

This general conception of vision and visual processing has had a predicable impact on related fields such as computer vision. The most influential theorist with respect to how visual percepts might be computed within the framework established by the classical physiological work of Hubel and Wiesel has been David Marr, the first person to explicitly describe vision as an information-processing task, and to give a detailed exposition of how this task might be implemented in the visual system (Marr, 1982). Marr proposed three stages of processing that entail the construction of the so-called "primal sketch", the "2.5-D sketch" and ultimately a "3-D model representation". In keeping with the known physiology of the day, the generation of the primal sketch was supposed to involve the detection of edges and other elementary image features. The construction of the 2.5-D sketch, in Marr's conception, depended on an extraction of the orientation and depth of surfaces delineated by contours; the 3-D representation of the physical world derived from the information extracted in these earlier steps would then correspond to what observers see. The idea that a final representation of the world is generated by an algorithmic processing of the retinal image, predicated on the detection of image features, is much the same concept of visual processing that stemmed from visual physiology. Indeed, Marr explicitly wanted his scheme to correspond with the physiology of the visual system.

A DIFFERENT INTERPRETATION BASED ON SPATIAL FREQUENCY

Despite the attractiveness and prevalence of this classical way of thinking about visual processing in the closing decades of the 20th C., at least one alternative view of the functional significance of visual cortical neurons emerged more or less in parallel with the conventional ideas about receptive fields summarized here. This position, initially held by a minority of vision scientists, is based on the idea of—and considerable evidence for—receptive fields as spatial frequency filters.

In this interpretation of visual receptive fields and their function, the early stages of visual processing are imagined to entail a large number of filters or "channels" that are "tuned" to different spatial frequencies. This line of thought, initiated by Fergus Campbell and John Robson in the late 1960s (see Campbell and Robson, 1968), is predicated on Fourier's theorem, which states that any continuous periodic function can be decomposed into a set of sinusoids and that the original function can therefore be reconstructed from this set of sinusoids, which is referred to as its Fourier spectrum. This sort of analysis had earlier been suggested as the way the basilar membrane of the inner ear extracts frequency components from complex sound signals, and, as Campbell and Robson recognized, the idea can be applied just as well to images. Thus any image can be decomposed into a Fourier spectrum that comprises a set of sinusoidal gratings with different spatial frequencies; and, by reversing the process, the image can be reconstructed from this spectrum.

This general conception of vision seemed promising in that it explained the otherwise puzzling findings reported in several psychophysical studies. For example, when human observers are presented with image patterns of alternating dark and light stripes, the minimal level of contrast between the stripes required to detect the pattern varies according to the spatial frequency of the stripes, a fact that defines the so-called "contrast sensitivity function" (Blakemore and Campbell, 1969; Graham and Nachmias, 1971). This result, together with evidence that at least some cells in the primary visual cortex are tuned to different spatial frequency ranges, supported the idea that the neurons conventionally referred to as "edge detectors" (see above) might actually be local spatial frequency analyzers, or "filters". The contrast sensitivity function demonstrated psychophysically would thus arise from the different contrast sensitivities of these filters. The purpose of the relevant neurons would be to encode the spatial frequency of the area of the image falling within their receptive fields, operating in the same general way as a Fourier analysis (Figure 9.2). (De Valois et al., 1982; De Valois and De Valois, 1988).

This idea was extended to moving stimuli by the further suggestion that motion selective neurons might function as spatio-temporal filters, thus analyzing the frequency

Implications

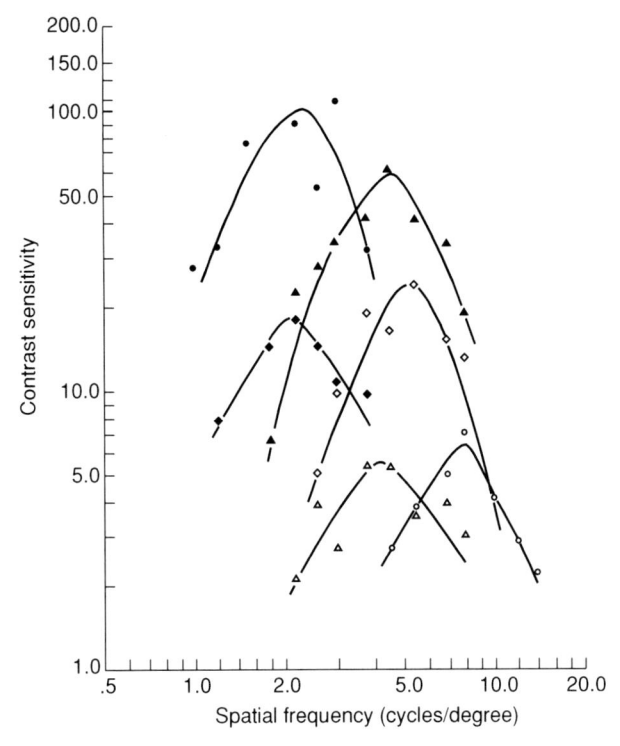

Figure 9.2 Example of the spatial frequency tuning of six cells monitored by electrophysiological recording in primary visual cortex of a monkey. The contrast sensitivity of each cell was measured by presenting alternating light and dark stripes at different spatial frequencies to the anesthetized animal; the minimum contrast of the stimuli required to make each cell fire above baseline rate was then plotted. Each of the six cells is most sensitive to a different range of spatial frequencies. (After De Valois et al., 1982)

structure of the retinal stimulus over time as well as in space (Adelson and Bergen, 1985; Watson and Ahumada, 1985; DeAngelis et al., 1993a, b). Much like the influence of physiological evidence on Marr's computational theory, evidence for spatio-temporal filtering has also had an impact on computer vision, where the output of various filters, rather than the output of edge detectors, has been used as a basis for modeling some aspects of higher level visual processing (see, for example, Heeger, 1988; Malik and Perona, 1990; Jones and Malik, 1992).

A PROBLEM WITH THESE INTERPRETATIONS OF VISUAL NEURONAL FUNCTION

Both the more traditional receptive field approach and spatial frequency theories of visual processing share the belief that the primary role of neurons in the visual cortex is to encode and represent retinal image features, whether as concrete characteristics such as a particular contour, or more abstract features such as the spatial frequency spectrum

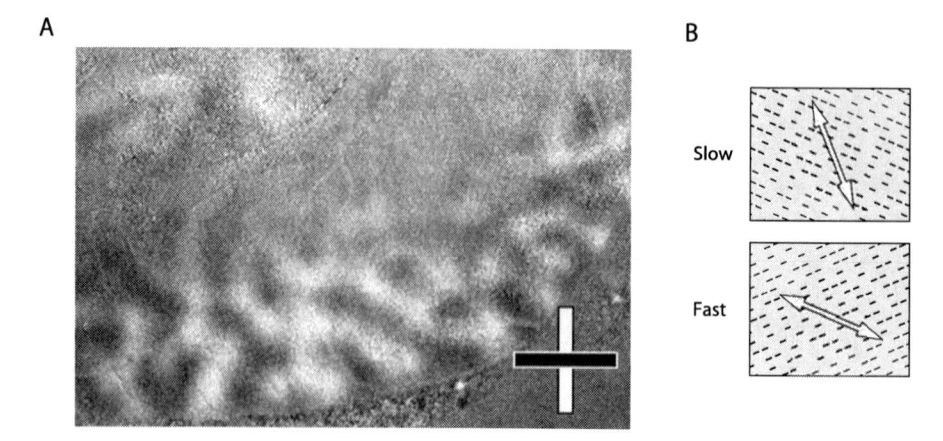

Figure 9.3 Evidence obtained by optical imaging showing that the same pattern of cortical activity in V1 can be elicited by different stimuli (the experimental animal in this case was a ferret). The optical imaging technique monitors cortical activity by virtue of activity-dependent changes in the light reflected from cortical surface layers at particular wavelengths (the dark areas are more active; the view is looking down on the surgically exposed primary visual cortex). (A) The same pattern of neuronal activity can be elicited by either of the two different stimuli in (B). (B) Examples of two stimuli comprising differently orientated line segments moving in different directions at different speeds, that elicited the same pattern of activity in the primary visual cortex shown in (A). (After Basole et al., 2003)

of a particular portion of an image. An implied corollary is that such representations correspond to, and will ultimately explain, visual percepts.

In addition to the evidence summarized in previous chapters that what we see (and thus visual processing) corresponds not to stimulus features but to the image-source statistics accumulated over the history of human experience, recent physiological observations have also begun to challenge the earlier consensus about feature detection and representation.

A key observation in this regard is that the activity of some visual cortical neurons cannot be understood in terms of their receptive field properties, at least as these properties have been conventionally defined. For example, David Fitzpatrick and his collaborators have shown that the *same* pattern of neuronal activity in V1 can be elicited by differently oriented stimuli moving in different directions at different speeds (Figure 9.3) (Basole et al., 2003). This result is contrary to what would be expected if the pattern of activity simply represented the combined neuronal selectivities for orientation, direction and speed. Although the finding illustrated in Figure 9.3 can be rationalized in terms of a spatio-temporal filtering model (as the authors suggest), it raises doubts about standard conceptions of receptive field properties and their relation to perception. Indeed, misgivings about mainstream thinking had already begun to be expressed by some physiologists (see, for example, Lennie, 1998).

Other recent observations have also challenged the conventional concept of receptive field properties by showing that the context of particular stimulus features modulates the relevant neuronal responses in a wide variety of ways. It is now generally recognized that the response properties of visual cortical neurons are influenced, often markedly, by stimuli presented outside the region of visual space that has traditionally

Implications

defined the extent of a neuron's receptive field (reviewed in Fitzpatrick, 2000; Worgotter and Eysel, 2000). For instance, the response of orientation-selective cells in V1 to a moving bar is suppressed in varying degrees by the presence of moving bars outside the receptive field, even though the neurons show no response when the stimulus outside the field is presented alone (Knierim and Van Essen, 1992). These findings are not particularly surprising, considering that neurons at different levels of the primary visual processing pathway receive a majority of their synaptic inputs from other neurons at the same level and/or feedback from neurons at higher levels of processing. They cast doubt, however, on the notion that a "representation" of the retinal image is in any sense reconstructed at some level of the visual system based on the combined receptive field properties of the relevant neurons. These countervailing observations about receptive fields should not be taken to mean that the evidence illustrated in Figure 9.1 is in any sense wrong. Rather, they imply that conventional thinking about the relationship between physiology and perception is at best incomplete.

The fundamental problem underlying this incompleteness is that the relationship between retinal stimuli and their physical sources is inevitably uncertain; as a result, the link between images and sources needed to ensure behavioral success is necessarily probabilistic. Thus, even if representations of retinal images could somehow be achieved through a combination of neuronal receptive field properties, there would be no logical way to link such representations to the real-world sources underlying the retinal stimuli. If rule-based algorithms cannot specify the relationship between a retinal stimulus and its source, there is no way to logically relate the receptive field properties of neurons as they are currently understood to the percepts that we actually see.

A PROBABILISTIC CONCEPTION OF VISUAL PROCESSING

These physiological observations, together with present evidence about the basis of perceived geometry and observations about the nature of several other aspects of visual perception (see Purves and Lotto, 2003 for a recent review), make it highly likely that the basic scheme of visual processing is a probabilistic one. If this concept of vision is correct, then neuronal activity cannot generate perceptions of scene geometry simply by encoding features in the retinal image. Given that the relation of geometrical features in the image plane to their physical sources is statistical, neuronal responses to any geometrical projection should reflect the probability distributions that describe the possible spatial arrangements in the physical world that could underlie the 2-D patterns in the retinal image plane. In this conception, the activity of a population of visual neurons is determined by the combined statistical influence of *all* the parameters in a given stimulus capable of conveying information about the world via the medium of light.

The strong contextual influences on the orientation selectivity of V1 cells imply, at least in broad terms, how this probabilistic scheme of processing might be expressed in neuronal physiology. For instance, the fact discussed earlier that neuronal responses to oriented lines are influenced by the orientation of objects outside their classical receptive fields suggests that the relevant neurons respond to the geometrical information in stimuli by generating a pattern of population activity that reflects the probability distribution of the possible sources of this particular aspect of the stimulus. Thus rather than encoding a specific feature such as line orientation, the ensuing neuronal activity would signify the relative frequency of occurrence of the possible physical sources related to the geometrical pattern in the stimulus (e.g., a line in the context of other lines in different orientations). There is, in this framework, no representation of the image in the relevant neural activity, but only a manifestation of the statistical linkage between the retinal image and its possible sources.

RELATION TO THE HIERARCHICAL ORGANIZATION OF THE VISUAL SYSTEM

How, then, does this probabilistic framework relate to the hierarchical organization of the visual pathway, a scheme that fits well with intuitions about feature representation?

V1 is a clearly hub in the primary visual pathway where information from the two lateral geniculate nuclei converges for the first time. At the same time, V1 receives an enormous amount of information back from other cortical regions, these connections being far more prevalent than synapses arising from retinal input via the geniculate nuclei. In a probabilistic processing framework, this arrangement suggests that V1 activity reflects the conjoint probability distribution of all the essential parameters of the possible physical sources of the stimulus. From this perspective, the activity of a neuron or a small group of neurons in V1 would essentially act as an estimator of the value at a point in the probability function.

Of course, this concept of V1 activity is complicated by the fact that a conjoint probability distribution pertinent to any natural stimulus has many dimensions: a large number of parameters are obviously required to describe the different categories of information germane to the possible physical sources of a stimulus (only a few geometrical features have been considered in earlier chapters). Such probability distributions are thus likely to be expressed in a marginal ("folded") form in V1. Given the large but nonetheless limited number of neurons in V1 (estimated to be at least 8×10^8), it seems likely that some parameters of the conjoint distribution are integrated to reduce the dimensionality. The further projections from V1 to different loci in the extrastriate visual cortices may therefore serve to "unfold" the probability distributions embodied in the primary visual cortex. If so, the activity of neurons in various extrastriate areas would correspond to different subsets of the conjoint probability distribution in fuller form, reflecting a particular aspect of the possible sources of visual stimuli, such as motion or color. This suggestion is consistent with the fact that the "higher-order" visual areas that have been most thoroughly investigated (e.g., MT and V4 in the primate cortex) appear to be responsible for a particular subset of the visual qualities that define the world we see (e.g., motion in MT or color in V4, or particularly important objects such as faces in the inferior temporal lobe).

EXAMPLES OF HOW SUCH A PROBABILISTIC STRATEGY OF VISION MIGHT OPERATE

This way of considering the functional significance of visual processing circuitry can certainly be applied to understanding the neuronal basis of some of the phenomena

Implications   **105**

addressed in earlier chapters. Consider, as perhaps the simplest example, the variation in the apparent length of a line as a function of its orientation (see Chapter 3). There has been little consensus about how, in general terms, the length of a line in the retinal stimulus might be represented by the properties of visual cortical neurons. Three hypotheses have been considered: 1) that length is encoded by "end-stopped" cells whose activity is suppressed by a line that extends beyond a certain limit (Hubel and Wiesel, 1962; Movshon et al., 1978); 2) that neurons tuned to different spatial frequencies distinguish stimulus length (De Valois et al., 1982; DeAngelis et al., 1994); and 3) that the number of active cells in response to a stimulus correlates with the length of the stimulus (an intuitive idea that has been discussed but not written about). None of these ideas, however, provides a convincing way to explain perceived line length. The first and the second hypotheses are flawed in that they would work only for stimuli that happen to fall within the classical receptive fields of the relevant neurons (most visual stimuli obviously do not). The third hypothesis is challenged by the finding that the extent of V1 activity in response to a given stimulus is influenced by luminance contrast, and thus does not correlate very well with spatial extent of the stimulus (Sceniak et al., 1999; see also Kapadia et al., 1999).

The failure to identify a clear physiological correlate of stimulus length is not particularly surprising given the evidence in Chapter 3 that apparent length is determined probabilistically. In a statistical framework, cortical activity does not encode the length of the stimulus per se, but is a manifestation of the conjoint probability distribution of all the possible physical sources of the stimulus. Although length may not be an independent dimension of the conjoint distribution, this parameter must nonetheless be fully embedded in it. The fact that the perceptual variation of line length as a function of orientation is so well predicted by the statistical linkage between projected length on the retina and the physical sources derived from natural scenes offers strong support for this sort of interpretation.

Another example is the misperception of angles. As noted, the complex range of interactions that has now been discovered among orientation-selective cells in V1 contradicts assumptions about the physiological basis of this effect being simply lateral inhibitory interactions among the V1 neurons responsive to the stimulus (see Chapter 4). These cortical interactions can, however, be understood as instantiating the full range of statistical relationships between the stimulus and its possible sources. As shown in Chapter 4, the angle formed by any two lines is shifted toward 90° on the pertinent empirical scale compared to the position of the angle in geometrical space. Thus a testable prediction is that the peaks of the overall cortical activity elicited by the two lines will be shifted toward orientation domains more orthogonal in their selectivity than the peaks of activity elicited by each line alone.

SUMMARY

In retrospect, neither the concept of classical receptive fields (now outmoded in any event), nor the concept of spatial frequency filtering, is able to reconcile present physiological and perceptual evidence pertinent to geometry. The observations described in the preceding chapters imply that the circuitry underlying this and other aspects of visual perception can best be understood in terms of the establishment and continuous enrichment of a probabilistic linkage between retinal images and their possible physical sources. The biological rationale for this physiology is a means of contending with the inverse optics problem, which we take to be the fundamental challenge in the evolution of biological vision. The probabilistic relationship between the geometrical characteristics of retinal stimuli and their real-world sources predicts virtually all the better known discrepancies between the measured properties of geometrical stimuli and the percepts they elicit. Given the historical difficulty rationalizing visual perception in terms of conventional receptive field properties, it makes sense to now explore the operation of visual processing circuitry in these terms.

References

Adelson EH, Bergen JR (1985) Spatiotemporal energy models for the perception of motion. Journal of the Optical Society of America A, Optics, Image Science & Vision 2:284–299.

Andrews DP (1967) Perception of contour orientation in the central fovea. I: short lines. Vision Research 7:975–997.

Atick JJ, Redlich AN (1992) What does the retina know about natural scenes? Neural Computation 4:196–210.

Barlow HB (1961) Possible principles underlying the transformation of sensory messages. In: Sensory Communication (Rosenblith WA, ed), pp 217–234. Cambridge, MA: MIT Press.

Barlow HB, Blakemore C, Pettigrew JD (1967) The neural mechanism of binocular depth discrimination. Journal of Physiology (London) 193:327–342.

Basole A, White LE, Fitzpatrick D (2003) Mapping multiple stimulus features in the population response of visual cortical neurons. Nature 423:986–990.

Baylis GC, Rolls ET, Leonard CM (1987) Functional subdivisions of the temporal lobe neocortex. Journal of Neuroscience 4:2051–2062.

Bell AJ, Sejnowski TJ (1997) The "independent components" of natural scenes are edge filters. Vision Research 37:3327–3338.

Besl PJ (1988) Active, optical range imaging sensors. Mach Vision Appl 1:127–152.

Blakemore C, Campbell FW (1969) On the existence of neurons in the human visual system selectively responsive to the orientation and size of retinal images. Journal of Physiology (London) 203:237–260.

Blakemore C, Tobin EA (1972) Lateral inhibition between orientation detectors in the cat's visual cortex. Experimental Brain Research 15:439–440.

Blakemore C, Carpenter RHS, Georgeson MA (1970) Lateral inhibition between orientation detectors in the human visual system. Nature 228:37–39.

Bolles RC (1969) The role of eye movements in the Müller-Lyer illusion. Perception & Psychophysics 6:175–176.

Bouma H, Andriessen JJ (1970) Induced changes in the perceived orientation of line segments. Vision Research 10:333–349.

Bross M, Blair R, Longtin P (1978) Assimilation theory, attentive fields, and the Mueller-Lyer illusion. Perception 7:297–304.

Bruce CJ, Desimone R, Gross CG (1981) Visual properties of neurons in a polysensory area in superior temporal sulcus of the macaque. Journal of Neurophysiology 46:369–384.

Burns BD, Pritchard R (1971) Geometrical illusions and the response of neurons in the cat's visual cortex to angle patterns. Journal of Physiology (London) 213:599–616.

Campbell FW, Robson JG (1968) Application of Fourier analysis to the visibility of gratings. Journal of Physiology 197:551–566.

Carpenter RHS, Blakemore C (1973) Interactions between orientations in human vision. Experimental Brain Research 18:287–303.
Carr HA (1935) An introduction to space perception. New York, NY: Longmans, Green.
Changizi MA, Widders D (2002) Latency correction explains the classical geometrical illusions. Perception 31:1241–1262.
Chiao CC, Cronin TW, Osorio D (2000) Color signals in natural scenes: Characteristics of reflectance spectra and effects of natural illuminants. Journal of the Optical Society of America A, Optics, Image Science & Vision 17:218–224.
Coren S (1971) A size-contrast illusion without physical size differences. American Journal of Psychology 84:565–566.
Coren S, Girgus JS (1978) Seeing is deceiving: The psychology of visual illusions. Potomac, MD: Lawrence Erlbaum.
Coren S, Enns JT (1993) Size contrast as a function of conceptual similarity between test and inducers. Perception & Psychophysics 54:579–588.
Cormack EO, Cormack RH (1974) Stimulus configuration and line orientation in the horizontal-vertical illusion. Perception & Psychophysics 16:208–212.
Craven BJ (1993) Orientation dependence of human line-length judgements matches statistical structure in real-world scenes. Proceedings of Royal Society of London Series B, Biological Sciences 253:101–106.
Day RH (1972) Visual spatial illusions: A general explanation. Science 175:1335–1340.
Day RH (1973) The Poggendorff illusion with obtuse and acute angles. Perception & Psychophysics 14:590–596.
Day RH, Dickinson RG (1976) The components of the Poggendorff illusion. British Journal of Psychology 67:537–552.
De Valois RL, De Valois KK (1988) Spatial Vision. New York: Oxford University Press.
De Valois RL, Albrecht DG, Thorell LG (1982) Spatial frequency selectivity of cells in macaque visual cortex. Vision Research 22:545–559.
DeAngelis GC, Ohzawa I, Freeman RD (1993a) Spatiotemporal organization of simple-cell receptive fields in the cat's striate cortex. I. General characteristics and postnatal development. Journal of Neurophysiology 69:1091–1117.
DeAngelis GC, Ohzawa I, Freeman RD (1993b) Spatiotemporal organization of simple-cell receptive fields in the cat's striate cortex. II. Linearity of temporal and spatial summation. Journal of Neurophysiology 69:1118–1135.
DeAngelis GC, Freeman RD, Ohzawa I (1994) Length and width tuning of neurons in the cat's primary visual cortex. Journal of Neurophysiology 71:347–374.
Delboeuf JLR (1892) 'Sur une nouvelle illusion d'optique'. Bulletin de l'Academie royale de Belgique 24:545–558.
Desimone R, Gross CG (1979) Visual areas in the temporal cortex of the macaque. Brain Research 178:363–380.
Dewar RE (1967) The effect of angle between the oblique lines on the decrement of the Müller-Lyer illusion with extended practice. Perception & Psychophysics 2:426–428.
Dixon MW, Proffitt DR (2002) Overestimation of heights in virtual reality is influenced more by perceived distal size than by the 2-D versus 3-D dimensionality of the display. Perception 31:103–112.
Dong DW, Atick JJ (1995) Statistics of natural time-varying images. Network: Computation in Neural Systems 6:345–358.
Dragoi V, Sur M (2000) Dynamic properties of recurrent inhibition in primary visual cortex: Contrast and orientation dependence of contextual effects. Journal of Neurophysiology 83:1019–1030.
Earlebacher A, Sekuler R (1969) Explanation of the Müller-Lyer illusion: Confusion theory examined. Journal of Experimental Psychology 80:462–467.
Epstein W, Landauer AA (1969) Size and distance judgments under reduced conditions of viewing. Perception & Psychophysics 6:269–272.
Evans CR, Marsden RP (1966) A study of the effect of perfect retinal stabilization of some well-known visual illusions, using the after-image as a method of compensating for eye movements. British Journal of Physiological Optics 23:242–248.
Field DJ (1994) What is the goal of sensory coding? Neural Computation 6:559–601.

References

Fisher GH (1969) An experimental study of angular subtension. Quarterly Journal of Experimental Psychology 21:356–366.

Fisher GH (1970) An experimental and theoretical appraisal of the perspective and size-constancy theories of illusions. Quarterly Journal of Experimental Psychology 22:631–652.

Fitzpatrick D (2000) Seeing beyond the receptive field in primary visual cortex. Current Opinion in Neurobiology 10:438–443.

Foley JM (1985) Binocular distance perception: Egocentric distance tasks. Journal of Experimental Psychology: Human Perception & Performance 11:133–149.

Ganz L (1966) Mechanism of the figural after-effects. Psychological Review 73:128–150.

Geisler WS, Perry JS, Super BJ, Gallogly DP (2001) Edge co-occurrence in natural images predicts contour grouping performance. Vision Research 41:711–724.

Gibson JJ (1950) The perception of the visual world. Boston, MA: Houghton Mifflin.

Gibson JJ (1966) The Senses Considered As Perceptual Systems. Boston, MA: Houghton Mifflin.

Gibson JJ (1979/1986) The Ecological Approach to Visual Perception. Hillsdale, New Jersey: Lawrence Erlbaum.

Gilbert CD, Wiesel TN (1990) The influence of contextual stimuli on the orientation selectivity of cells in primary visual cortex of the cat. Vision Research 30:1689–1701.

Gillam B (1971) A depth processing theory of the Poggendorff illusion. Perception & Psychophysics 10:211–216.

Gillam B (1995) The perception of spatial layout from static optical information. In: Perception of space and motion (Epstein W, Rogers SJ, eds), pp 23–67. San Diego: Academic Press.

Gillam B (1998) Illusions at century's end. In: Perception and Cognition at Century's End, 2nd Edition (Hochberg J, ed), pp 98–136. New York: Academic Press.

Girgus JS, Coren S (1975) Depth cues and constancy scaling on the horizontal-vertical illusion: the bisection error. Canadian Journal of Psychology 29:59–65.

Girgus JS, Coren S, Agdern M (1972) The interrelationship between the Ebbinghaus and Delboeuf illusions. Journal of Experimental Psychology 95:453–455.

Gogel WC (1965) Equidistance tendency and its consequences. Psychological Bulletin 64:153–163.

Gogel WC, Tietz JD (1979) A comparison of oculomotor and motion parallax cues of egocentric distance. Vision Research 19:1161–1170.

Graham N, Nachmias J (1971) Detection of grating patterns containing two spatial frequencies: A comparison of single-channel and multiple-channels models. Vision Research 11:251–259.

Green RT, Stacey BG (1966) Misapplication of the misapplied constancy hypothesis. Life Sciences 5:1871–1880.

Greene E (1994) Collinearity judgment as a function of induction angle. Perceptual & Motor Skills 78:655–674.

Gregory RL (1963) Distortion of visual space as inappropriate constancy scaling. Nature 199:678–780.

Gregory RL (1966) Eye and Brain: The Psychology of Seeing. New York: McGraw Hill.

Gregory RL (1968) Perceptual illusions and brain models. Proceedings of Royal Society of London Series B, Biological Sciences 171:279–296.

Gregory RL (1974) Concepts and Mechanisms of Perception. London: Duckworth.

Gregory RL (1997) Eye and Brain. Princeton, NJ: Princeton University Press.

Griggs R (1974) Constancy scaling theory and the Mueller-Lyer illusion: More disconfirming evidence. Bulletin of the Psychonomic Society 4:168–170.

Hartline HK (1969) Visual receptors and retinal interaction. Science 164:270–278.

Heeger DJ (1988) Optical flow using spatio-temporal filters. International Journal of Computer Science 1:279–302.

Helmholtz HLFv (1866/1924) Helmholtz's Treatise on Physiological Optics, Third German Edition Edition. Rochester, NY: The Optical Society of America.

Helson H (1964) Adaptation-level theory. New York: Harper & Row.

Hering E (1861) 'Beiträge zur Physiologie'. Leipzig: Engelman.

Hershenson M (1999) Visual Space Perception. Cambridge, MA: The MIT Press.

Heywood S, Chessell K (1977) Expanding angles? Systematic distortions of space involving angular figures. Perception 6:571–582.

Higashiyama A (1996) Horizontal and vertical distance perception: The discorded-orientation theory. Perception & Psychophysics 58:259–270.
Hotopf WH, Robertson SH (1975) The regression to right angles tendency, lateral inhibition, and the transversals in the Zollner and Poggendorff illusions. Perception & Psychophysics 18:453–459.
Howard RB, Wagner M, Mills RC (1973) The superiority of the pair-comparisons method for scaling visual illusions. Perception & Psychophysics 13:507–512.
Howe CQ, Purves D (2002) Range image statistics can explain the anomalous perception of length. Proceedings of the National Academy of Sciences of the United States of America 99:13184–13188.
Howe CQ, Purves D (2004) Size contrast and assimilation explained by the statistics of natural scene geometry. Journal of Cognitive Neuroscience 16:90–102.
Howe CQ, Purves D (2005a) Natural scene geometry predicts the perception of angles and line orientation. Proceedings of the National Academy of Sciences of the United States of America 102:1228–1233.
Howe CQ, Purves D (2005b) The Müller-Lyer illusion explained by the statistics of image-source relationships. Proceedings of the National Academy of Sciences of the United States of America 102:1234–1239.
Howe CQ, Yang Z, Purves D (2005) The Poggendorff illusion explained by natural scene statistics. Proceedings of the National Academy of Sciences of the United States of America (in press).
Hoyer PO, Hyvärinen A (2000) Independent component analysis applied to feature extraction from colour and stereo images. Network: Computation in Neural Systems 11:191–210.
Hubel DH (1982) Exploration of the primary visual cortex, 1955–78. Nature 299:515–524.
Hubel DH, Wiesel TN (1959) Receptive fields of single neurons in the cat's striate cortex. Journal of Physiology (London) 148:574–591.
Hubel DH, Wiesel TN (1962) Receptive fields, binocular interaction and functional architecture in the cat's visual cortex. Journal of Physiology (London) 160:106–154.
Hubel DH, Wiesel TN (1968) Receptive fields and functional architecture of monkey striate cortex. Journal of Physiology (London) 195:215–243.
Hubel DH, Wiesel TN (1974) Sequence regularity and geometry of orientation columns in the monkey striate cortex. Journal of Comparative Neurology 158:267–294.
Hubel DH, Wiesel TN (1977) Ferrier Lecture: Functional architecture of macaque monkey visual cortex. Proceedings of Royal Society of London Series B, Biological Sciences 198:1–59.
Ikeda H, Obonai T (1955) Figural after-effect, retroactive effect and simultaneous illusion. Japanese Journal of Psychology 26:235–246.
Indow T (1991) A critical review of Luneburg's model with regard to global structure of visual space. Psychological Review 98:430–453.
Jaeger T, Pollack RH (1977) Effect of contrast level and temporal order on the Ebbinghaus circles illusion. Perception & Psychophysics 21:83–87.
Jaeger T, Lorden R (1980) Delboeuf illusions: Contour or size detector interactions? Perceptual & Motor Skills 50:376–378.
Jaeger T, Grasso K (1993) Contour lightness and separation effects in the Ebbinghaus illusion. Perceptual & Motor Skills 76:255–258.
Jones DG, Malik J (1992) Computational framework for determining stereo correspondence from a set of linear spatial filters. Image & Vision Computing 10:699–708.
Jordan K, Uhlarik J (1986) Length contrast in the Mueller-Lyer figure: Functional equivalence of temporal and spatial separation. Perception & Psychophysics 39:267–274.
Kapadia MK, Westheimer G, Gilbert CD (1999) Dynamics of spatial summation in primary visual cortex of alert monkeys. Proceedings of the National Academy of Sciences of the United States of America 96:12073–12078.
Knierim JJ, Van Essen DC (1992) Visual cortex: cartography, connectivity, and concurrent processing. Current Opinion in Neurobiology 2:150–155.
Kuennapas TM (1957) The vertical-horizontal illusion and the visual field. Journal of Experimental Psychology 53:405–407.
Kuffler SW (1953) Discharge patterns and functional organization of mammalian retina. Journal of Neurophysiology 16:37–68.

References

Leibowitz H, Toffey S (1966) The effect of rotation and tilt on the magnitude of the Poggendorff illusion. Vision Research 6:101–103.

Lennie P (1998) Single units and visual cortical organization. Perception 27:889–935.

Lewis EO (1909) Confluxion and contrast effects in the Mueller-Lyer illusion. British Journal of Psychology 3:21–41.

Li CY, Li W (1994) Extensive integration field beyond the classical receptive field of cat's striate cortical neurons—classification and tuning properties. Vision Research 34:2337–2355.

Loomis JM, Da Silva JA, Philbeck JW, Fukusima SS (1996) Visual perception of location and distance. Current Directions in Psychological Science 5:72–77.

Luckiesh M (1922) Visual illusions. Their causes, Characteristics and Applications. New York: D. Van Nostrand Company.

Luneburg RK (1947) Mathematical analysis of binocular vision. Princeton, NJ: Princeton University Press.

Maclean IE, Stacey BG (1971) Judgment of angle size: An experimental appraisal. Perception & Psychophysics 9:499–504.

Malik J, Perona P (1990) Preattentive texture discrimination with early vision mechanisms. Journal of the Optical Society of America A, Optics, Image Science & Vision 7:923–932.

Marr D (1982) Vision: A Computational Investigation into Human Representation and Processing of Visual Information. San Francisco: Freeman.

Massaro DW, Anderson NH (1971) Judgmental model of the Ebbinghaus illusion. Journal of Experimental Psychology 89:147–151.

Maunsell JHR, Van Essen DC (1983) Functional properties of neurons in middle temporal visual area of the macaque monkey. I. Selectivity for stimulus direction, speed, and orientation. Journal of Neurophysiology 49:1127–1147.

McManus IC (1978) The horizontal-vertical illusion and the square. British Journal of Psychology 69:369–370.

Minnaert MGJ (1937/1992) Light and color in the outdoors. New York: Springer.

Morrison JD, Whiteside TC (1984) Binocular cues in the perception of distance of a point source of light. Perception 13:555–566.

Morrison LC (1977) Inappropriate constancy scaling as a factor in the Muller-Lyer illusion. British Journal of Psychology 68:23–27.

Movshon JA (1975) The velocity tuning of single units in cat striate cortex. Journal of Physiology 249:445–468.

Movshon JA, Thompson ID, Tolhurst DJ (1978) Receptive field organization of complex cells in the cat's striate cortex. Journal of Physiology 283:79–99.

Movshon JA, Adelson EH, Gizzi MS, Newsome WT (1986) The analysis of moving visual patterns. In: Pattern Recognition Mechanisms (Chagas C, Gattass R, Gross C, eds), pp 148–163. New York: Springer-Verlag.

Müller-Lyer FC (1889) 'Optische Urteilstäuschungen'. Archiv für Anatomie und Physiologie Supplement-Band:263–270.

Nakagawa D (1958) Mueller-Lyer illusion and retinal induction. Psychologia 1:167–174.

Nelson JI, Frost BJ (1978) Orientation-selective inhibition from beyond the classic visual receptive field. Brain Research 139:359–365.

Nundy S, Lotto B, Coppola D, Shimpi A, Purves D (2000) Why are angles misperceived? Proceedings of the National Academy of Sciences of the United States of America 97:5592–5597.

Obonai T (1954) Induction effects in estimates of extent. Journal of Experimental Psychology 47:57–60.

Olshausen BA, Field DJ (1996) Emergence of simple-cell receptive field properties by learning a sparse code for natural images. Nature 381:607–609.

Olshausen BA, Field DJ (1997) Sparse coding with an overcomplete basis set: a strategy employed by V1? Vision Research 37:3311–3325.

Ooi TL, Wu B, He ZJ (2001) Distance determined by the angular declination below the horizon. Nature 414:197–200.

Oppel JJ (1855) 'Über geometrisch-optische Täuschungen'. Jahresbericht des Frankfurter Vereins 1854–1855, 55:37–47.

Owens DA, Leibowitz HW (1976) Oculomotor adjustments in darkness and the specific distance tendency. Perception & Psychophysics 20:2–9.

Oyama T (1960) Japanese studies on the so-called geometrical-optical illusions. Psychologia 3:7–20.
Parkhurst DJ, Niebur E (2003) Scene content selected by active vision. Spatial Vision 16:125–154.
Pearce D, Matin L (1969) Variation of the magnitude of the horizontal-vertical illusion with retinal eccentricity. Perception & Psychophysics 6:241–243.
Philbeck JW, Loomis JM (1997) Comparison of two indicators of perceived egocentric distance under full-cue and reduced-cue conditions. Journal of Experimental Psychology: Human Perception & Performance 23:72–85.
Poggio GF, Gonzalez F, Krause F (1988) Stereoscopic mechanisms in monkey visual cortex: binocular correlation and disparity selectivity. Journal of Neuroscience 8:4531–4550.
Pollock WT, Chapanis A (1952) The apparent length of a line as a function of its inclination. Quarterly Journal of Experimental Psychology 4:170–178.
Ponzo M (1928) 'Urteilstäuschungen über Mengen'. Archiv für die gesamte Psychologie 65.
Pressey AW (1967) A theory of the Mueller-Lyer illusion. Perceptual & Motor Skills 25:569–572.
Pressey AW (1970) The assimilation theory applied to a modification of the Müller-Lyer illusion. Perception & Psychophysics 8:411–412.
Pressey AW (1977) Measuring the Titchener circles and Delboeuf illusions with the method of adjustment. Bulletin of the Psychonomic Society 10:118–120.
Prinzmetal W, Gettleman L (1993) Vertical-horizontal illusion: One eye is better than two. Perception & Psychophysics 53:81–88.
Purves D, Lichtman JW (1985) Principles of Neural Development. Sunderland, MA: Sinauer.
Purves D, Lotto B (2003) Why We See What We Do: An Empirical Theory of Vision. Sunderland, MA: Sinauer.
Recanzone GH, Wurtz RH, Schwarz U (1997) Responses of MT and MST neurons to one and two moving objects in the receptive field. Journal of Neurophysiology 78:2904–2915.
Restle F, Merryman CT (1968) An adaptation-level theory account of a relative-size illusion. Psychonomic Science 12:229–230.
Robinson JO (1998) The Psychology of Visual Illusion. Mineola, NY: Dover Publications.
Rock I (1995) Perception. New York, NY: Scientific American Library.
Rolls ET (1984) Neurons in the cortex of the temporal lobe and in the amygdala of the monkey with responses selective for faces. Human Neurobiology 3:209–222.
Rolls ET (1994) Brain mechanisms for invariant visual recognition and learning. Behavioural Processes 33:113–138.
Ruderman DL, Bialek W (1994) Statistics of natural images: Scaling in the woods. Physical Review Letters 73:814–817.
Sceniak MP, Ringach DL, Hawken MJ, Shapley R (1999) Contrast's effect on spatial summation by macaque V1 neurons. Nature Neuroscience 2:733–739.
Schiffman HR, Thompson JG (1975) The role of figure orientation and apparent depth in the perception of the horizontal-vertical illusion. Perception 4:79–83.
Schwartz O, Simoncelli EP (2001) Natural signal statistics and sensory gain control. Nature Neuroscience 4:819–825.
Seckel A (2000) The Art of Optical Illusions. Carlton Books.
Sedgwick HA (1986) Space perception. In: Handbook of Perception and Human Performance: Vol. 1. Sensory processes and perception (Boff KR, Kaufman L, Thomas JP, eds), pp 21.21–21.57. New York: Wiley.
Shipley WC, Mann BM, Penfield MJ (1949) The apparent length of tilted lines. Journal of Experimental Psychology 39:548–551.
Simoncelli EP, Olshausen BA (2001) Natural image statistics and neural representation. Annual Review of Neuroscience 24:1193–1216.
Skottun BC (2000) Amplitude and phase in the Mueller-Lyer illusion. Perception 29:201–209.
Sleight RB, Austin TR (1952) The horizontal-vertical illusion in plane geometric figures. Journal of Psychology 33:279–287.
Tanaka K (1993) Neuronal mechanisms of object recognition. Science 262:685–688.
Thiéry A (1896) 'Über geometrisch-optische Taüschungen'. Philosophische Studien 12:67–126.
Thorell LG, De Valois RL, Albrecht DG (1984) Spatial mapping of monkey V1 cells with pure color and luminance stimuli. Vision Research 24:751–769.

References

Todd JT, Oomes AH, Koenderink JJ, Kappers AM (2001) On the affine structure of perceptual space. Psychological Science 12:191–196.

Tootell RB, Silverman MS, Hamilton SL, De Valois RL, Switkes E (1988) Functional anatomy of macaque striate cortex. III. Color. Journal of Neuroscience 8:1569–1593.

Turiel A, Parga N, Ruderman DL, Cronin TW (2001) Multiscaling and information content of natural color images. Physical Review E 62:1138–1148.

Van Essen DC, Gallant JL (1994) Neural mechanisms of form and motion processing in the primate visual system. Neuron 13:1–10.

Van Hateren JH, Ruderman DL (1998) Independent component analysis of natural image sequences yields spatio-temporal filters similar to simple cells in primary visual cortex. Proceedings of Royal Society of London Series B, Biological Sciences 265:2315–2320.

Van Hateren JH, van der Schaaf A (1998) Independent component filters of natural images compared with simple cells in primary visual cortex. Proceedings of Royal Society of London Series B, Biological Sciences 265:359–366.

von Collani G (1985) The horizontal-vertical illusion in photographs of concrete scenes with and without depth information. Perceptual & Motor Skills 61:523–531.

Wachtler T, Lee TW, Sejnowski TJ (2001) Chromatic structure of natural scenes. Journal of the Optical Society of America A, Optics, Image Science & Vision 18:65–77.

Wagner M (1985) The metric of visual space. Perception & Psychophysics 38:483–495.

Wallach H, O'Leary A (1982) Slope of regard as a distance cue. Perception & Psychophysics 31:145–148.

Watson AB, Ahumada AJ, Jr. (1985) Model of human visual-motion sensing. Journal of the Optical Society of America A, Optics Image Science & Vision 2:322–341.

Weintraub DJ, Krantz DH (1971) The Poggendorff illusion: amputations, rotations, and other perturbations. Perception & Psychophysics 10:257–264.

Wenderoth P, Parkinson A, White D (1979) A comparison of visual tilt illusions measured by the techniques of vertical setting, parallel matching, and dot alignment. Perception 8:47–57.

Worgotter F, Eysel UT (2000) Context, state and the receptive fields of striatal cortex cells. Trends in Neuroscience 23:497–503.

Wundt W (1862) 'Beiträge zur Theorie der Sinneswahrnehmung'. Leipzig und Heidelberg: C. F. Winter'sche Verlagshandlung.

Yang Z, Purves D (2003) A statistical explanation of visual space. Nature Neuroscience 6:632–640.

Yarbus AL (1967) Eye Movements and Vision. New York: Plenum Press.

Zeki SM (1983a) Colour coding in the cerebral cortex: The reaction of cells in monkey visual cortex to wavelengths and colours. Neuroscience 9:741–765.

Zeki SM (1983b) Colour coding in the cerebral cortex: The responses of wavelength-selective and colour-coded cells in monkey visual cortex to changes in wavelength composition. Neuroscience 9:767–781.

Zigler E (1960) Size estimates of circles as a function of size of adjacent circles. Perceptual & Motor Skills 11:47–53.

Glossary

aerial perspective The diminution of contrast (i.e., the increasing haziness of contour boundaries) as a function of distance from the observer; occurs as a result of the imperfect transmittance of the atmosphere, and is a monocular cue to depth.

algorithm A set of rules or procedures set down in logical notation, and typically (but not necessarily) carried out by a computer.

apparent Referring to what is perceived. For example, the apparent length of a line is how long the line appears to an observer.

artificial neural network A computer architecture for solving problems by feedback from trial and error, rather than by a predetermined algorithm.

assimilation Tendency for the perception of a target to include or be influenced by the characteristics of the background.

background Referring to the part or parts of a scene that are further away from an observer and/or less salient.

binocular disparity (See retinal disparity.)

binocular Referring to both eyes.

bottom-up A much-abused term that loosely refers to the flow of information from sensory receptors toward the cerebral cortex.

bottom-up processing Processing information according to the dictates of sensory receptors and input circuitry as such.

brightness Technically, the apparent intensity of a source of light; more generally, a sense of the effective overall intensity of a light stimulus (see lightness).

cerebral cortex The superficial gray matter covering the cerebral hemispheres.

cerebrum The largest and most rostral part of the brain in humans and other mammals, consisting of two cerebral hemispheres.

circuitry A general term in neurobiology referring to the connections between neurons; usually pertinent to some particular function (as in 'visual circuitry').

cognition A general term referring to 'higher order' mental processes; so vague in most usages as to have little substantive meaning.

color The subjective sensations elicited in humans (and presumably many other animals) by different spectral distributions of light.

context General term referring to the information provided by the surroundings of a 'target'. The division of a scene into target and surround is useful, but arbitrary, since *any* part of a scene provides contextual information for any other part.

contour A line or edge.

contrast The physical difference between the luminance (or spectral distribution in the case of color) of two surfaces. More specifically, the luminance difference or spectral difference between two regions of visual space, measured in percent (100% being the difference between low-reflecting and highly reflecting standards). Formally, $L_{max} - L_{min}/L_{max} + L_{min}$, which is called Michelson contrast.

cortex The gray matter of the cerebral hemispheres and cerebellum, where most of the neurons in the brain are located.

cumulative probability The summed probability of some variable being less than or equal to a particular value.

cumulative probability function Cumulative probability of a variable as a function of all the possible values of that variable.

degree Unit used to measure visual space based on the division a circle into 360°; 1 degree is approximately the width of the thumbnail held at arms length, and covers about 0.2 mm on the retina.

detector In vision, a nerve cell or other device that nominally detects the presence of some particular feature of visual stimuli (e.g., luminance, orientation etc.).

direction The course taken by something, e.g., a point, in a frame of reference; together with speed, defines velocity.

dorsal lateral geniculate nucleus The thalamic nucleus that relays information from the retina to the cerebral cortex; usually referred to as the lateral geniculate, or just the geniculate.

eccentricity Away from the center; in vision, refers to the distance in degrees away from the line of sight.

empirical Derived on the basis of past experience, effectively by trial and error.

empirical significance In the present context, what an individual observer or species has typically discovered a visual stimulus to be when interacting with stimulus sources in the environment.

Glossary

epiphenomenon An effect taken to be an incidental consequence of some more basic property or principle.

Euclidean space The three-dimensional space of conventional geometry.

extrastriate Referring to the regions of visual cortex that lie outside the primary (or striate) visual cortex.

extrastriate visual areas See extrastriate; includes areas such as V4, MT, and MST, which are taken to be particularly pertinent to the processing of one or another categories of visual information (e.g., color in V4, motion in MT and MST).

Fourier analysis (or transform) Mathematical procedure for representing any periodic function as the sum of a set of sinusoids.

Fourier's theorem The proposition proven by Jean-Batiste Fourier in the late 18th C. that any periodic function can be decomposed into a series of sine or cosine waves.

fovea Area of the human retina specialized for high acuity; contains a high density of cones and few rods. Most mammals do not have a well defined fovea, although many have an area of central vision (called the area centralis) in which acuity is higher than in more eccentric retinal regions.

frequency How often something occurs within a unit of time or space.

frequency distribution The relative frequency of occurrence of the possible values of a variable.

fronto-parallel plane Any plane orthogonal to the line of sight.

geometrical illusions Discrepancies between a visual stimulus and the resulting percept based on geometrical measurements (i.e., measurements of length, angle etc.).

gray matter General term describing regions of the central nervous system rich in neuronal cell bodies; includes the cerebral and cerebellar cortices, the nuclei of the brain, and the central portion of the spinal cord.

Hering illusion A classical geometrical effect in which parallel lines placed on a background of radiating lines look bowed.

heuristic A rule or procedure derived from past experience that can be used to solve a problem; in vision, such rules are sometimes taken to be the determinants of perception.

hierarchy A system of higher and lower ranks; in vision, the idea that neurons at lower stages of the visual system determine the properties of higher order neurons.

higher-order Processes and/or areas taken to be further removed from the input stages of a system; in neuroscience, this phrase is sometimes used as a synonym for cognitive processes.

higher-order neurons Neurons that are relatively remote from peripheral sensory receptors or motor effectors.

illuminant A source of illumination.

illumination The light that falls on a scene or surface.

illusions An imprecise word that refers to discrepancies between the physically measured properties of a visual stimulus and what is actually seen.

image formation The result of focusing the light rays diverging from a collection of adjacent points on object surfaces onto another surface (e.g., a screen or the retina) to form a corresponding set of points that represents the three dimensional sources on a two-dimensional plane.

image processing Improving (or otherwise changing) images by application of one or more algorithms.

image The representation of physical objects on a two-dimensional plane.

information The systematic arrangement of a parameter such that an observer (or a receiver) can , in principle, extract a signal from the background noise.

inhibitory response (inhibition) A cellular response involving a decrease in the rate of neural activity (action potentials generated per unit time).

input The information supplied to a neural processing system.

integration In neuroscience, the summation of excitatory and inhibitory synaptic conductance changes by postsynaptic cells.

lateral geniculate nucleus (See dorsal lateral geniculate nucleus.)

lateral inhibition Inhibitory effects extending laterally in the plane of neural tissue, e.g., the retina or the visual cortex; widely assumed to play a major role in perceptual phenomenology.

learning The acquisition of novel information or a new behavior through experience.

light The range in the electromagnetic spectrum that elicits visual sensations in humans (wavelengths of about 400–700 nm).

lightness The apparent reflectance of a surface (or transmittance of a medium), usually considered in terms of achromatic percepts ranging from white through grays to black (see brightness).

line The geometrical concept of a one-dimensional (straight or curved) entity.

line of sight An imaginary straight line from the center of the fovea through the point of fixation in visual space.

linear perspective The geometrical changes that arise when light reflected from three-dimensional objects is projected onto a two-dimensional surface.

Glossary

luminance The physical (photometric) intensity of light returned to the eye (or some other detector) adjusted for the sensitivity of the average human observer.

marginal probability distribution Probability distribution of a variable obtained from conjoint probability distribution of several variables by integrating out the other variables.

medium In the context of vision, a substance (e.g., the atmosphere or a filter) interposed between an observer and the objects in a scene.

modality A category of function; for example, vision, hearing, and touch are different sensory modalities.

monocular cues Term used to describe information (often about depth) arising from the view of a single eye.

motion The changing position of an object defined by speed and direction in a frame of reference.

motion parallax The different relative apparent movement of near and far objects as a function of moving the head or body while observing a scene.

Müller-Lyer illusion A geometrical effect in which the length of a line terminated by arrowheads appears shorter than the same line terminated by arrow tails; first described by the 19th C. German philosopher and sociologist F. D. Müller-Lyer.

nerve A collection of peripheral axons that are bundled together and travel a common route.

neural processing A general term used to describe the operations carried out by neural circuitry.

neuron Cell specialized for the conduction and transmission of electrical signals in the nervous system.

objects The physical entities that give rise to visual stimuli by reflecting illumination (or by emitting light, if, as more rarely happens, they are themselves generators of light energy).

occipital cortex Part of the cerebral cortex nearest the back of the head, containing mainly visual processing areas.

occipital lobe The posterior of the four lobes of the human cerebral hemisphere; primarily devoted to vision.

occlusion Blockage of objects in a visual scene by an object closer to the observer.

ontogeny The developmental history of an individual animal; used as a synonym for development.

optic nerve The nerve (cranial nerve II) containing the axons of retinal ganglion cells; extends from the eye to the optic chiasm.

optic tract The axons of retinal ganglion cells after they have passed through the region of the optic chiasm en route to the lateral geniculate nucleus of the thalamus.

orientation The arrangement of an object in the three dimensions of Euclidean space.

orientation selectivity Describing neurons that respond selectively to edges presented over a relatively narrow range of stimulus orientations.

orientation tuning curve The function obtained when a neuron's receptive field is tested with stimuli at the different orientations.

orthogonal Making a right angle with another line or surface; also called 'normal'.

parallel processing The simultaneous processing of visual or other information by different components or pathways in a sensory (or other) system.

perception The subjective awareness (typically taken to be conscious awareness) of any aspect of the external or internal environment.

perspective In art or other forms of graphical representation, any of various techniques for representing three-dimensional objects and depth relationships on a two-dimensional surface.

phenomenology General word used to describe the behavior of something (such as the phenomenology of visual perceptions).

photoreceptors Cells in the retina specialized to absorb photons, and thus to generate neural signals in response to light stimuli.

phylogeny The evolutionary history of a species or other taxonomic category.

pixel Member of the array of discrete elements that comprises a digital image.

Poggendorff illusion A geometrical effect that entails seeing an angled, collinear line that is occluded as being non-collinear; first described in the mid-19th C. by J. C. Poggendorff.

point The geometrical concept of a dimensionless location in space.

Ponzo illusion A geometrical effect that entails seeing two identical horizontal lines as being unequal in length when they are placed between two converging lines; first described in the early 20th C. by Italian psychologist M. Ponzo.

primary sensory cortex Any one of several cortical areas in direct receipt of the thalamic input for a particular sensory modality.

primary visual cortex (See striate cortex.)

primary visual pathway (retino-geniculocortical pathway) Pathway from the retina via the lateral geniculate nucleus of the thalamus to the primary visual cortex; carries the information that allows conscious visual perception.

primate Order of mammals that includes lemurs, tarsiers, marmosets, monkeys, apes, and humans (technically, a member of this order).

Glossary

probability The likelihood of an event, usually expressed as a value from 0 (will never occur) to 1 (will always occur).

probability distribution Probability of a variable having a particular value, expressed as a function of all the possible values of that variable.

psychology The study of mental processes in humans and other animals.

psychophysics The study of mental processes by quantitative methods, typically by reports from human subjects of the sensations elicited by carefully controlled stimuli.

radiance The electromagnetic energy emitted by an object.

range The distance of a point in space from an observer or a measuring device.

range image A digital image that includes information about the range (i.e., distance) of every constituent pixel.

ray Term used to indicate the passage of photons from source to a target or detector in a straight line.

real-world Phrase used to convey the idea that there is an external world that determines what we see, even though that world is directly unknowable.

receptive field properties The response characteristics of a neuron, defined by the region of the body surface (e.g., the region of the retina) and the stimulus qualities (e.g., orientation, length) that cause the neuron to alter its baseline activity.

receptor Nerve cells specialized for the transduction of physical energy into neural signals.

reflectance The percentage of incident light reflected from a surface (often expressed as the reflectance efficiency function, in which the reflectance of a surface is measured at different wavelengths).

reflection The return of light from a surface as a result of the incident light failing to be either absorbed or transmitted.

refraction The altered direction and speed of light as a result of passing from one medium to another (e.g., from air to the substance of the cornea).

representation In vision, the idea that the visual system reconstructs the retinal image, either literally or figuratively for 'presentation' a second time in the visual cortex. More generally, the transformation of information into another form or medium.

resolution In vision, the ability to distinguish two nearby points in space.

retina Laminated neural component of the eye that contains the photoreceptors (rods and cones) and the initial processing circuitry for vision.

retinal disparity The geometrical difference between the same points in the images simultaneously projected on the two retinas, measured in degrees with respect to the center of the fovea.

retinal ganglion cells The output neurons of the retina, whose axons form the optic nerve.

retinal image The image focused on the retina by the optical properties of the eye.

retinotopic map A map in which neighbor relationships at the level of the retina are maintained at higher stations in the visual system.

retinotopy The maintenance of the neighbor relationships at progressively higher stations in the visual system.

rotation A physical movement defined by an angular change in the position of a point or points in a frame of reference.

scale An ordering of quantities according to their magnitudes.

scatter Dispersion of light that degrades an image.

scene The real-world arrangement of objects and illumination with respect to the observer that gives rise to visual stimuli ('source' is a synonym).

sensation The subjective experience elicited by energy impinging on an organism's sensory receptors (a word that should be regarded with suspicion when used to differentiate this experience from 'perception').

sensitivity The degree of ability to respond to the energy in a sensory stimulus.

sensory Pertaining to sensation.

sensory system Term used to describe all the components of the central and peripheral nervous system concerned with sensation (or a particular modality such as vision).

shadows Regions of diminished light that occur when objects are interposed between light sources such as the sun and a surface potentially in receipt of that light.

sinusoid Pattern defined by a sine (or cosine) function.

software The programs that run computers.

spatial frequency The spatial interval over which a pattern repeats, usually measured in cycles/degree (or cycles/mm). More specifically, the number of cycles of luminance variation by some measure in a given direction over $1°$ of visual angle.

species A taxonomic category subordinate to genus; members of a species are defined by extensive similarities and the ability to interbreed.

spectral differences Differences in the distribution of spectral power in a visual stimulus that give rise to perceptions of color.

spectrum (pl. spectra) The power distribution of a given light source at different wavelengths; more generally, the distribution of a continuous variable.

speed The rate of change that, together with direction, defines the velocity of a moving object.

striate cortex The primary visual cortex in the occipital lobe in humans and other primates (also called Brodmann's area 17 or V1). So named because the prominence of layer IV in myelin-stained sections gives this region a striped appearance.

surface Any physical interface separable from the medium in which it resides.

synapse Specialized apposition between a neuron and a target cell; typically transmits information by release and reception of a chemical transmitter agent.

target An arbitrarily selected portion of a visual image whose perception is to be assessed.

texture The pattern of variation in the intensity of light reflected from a surface.

thalamus A collection of nuclei that forms the major component of the diencephalon. Although its functions are many, a primary role of the thalamus is to relay sensory information from the periphery to the cerebral cortex.

threshold The lowest energy level of a stimulus that causes a perceptual response; also, the level of membrane potential at which an action potential is generated.

T-illusion A geometrical effect in which a vertically oriented line looks longer than a horizontally oriented line of the same length.

top-down A much abused term that refers to the effects of what are taken to be 'higher order' mental processes on primary sensory or other 'bottom-up' information.

top-down processing The idea that cognitive influences arising from 'higher order' cortical regions influence 'lower order' cortical or sub-cortical processing.

transmittance The amount of light that reaches the eye or some other detector from a surface, compared to the amount that is initially reflected from it (or emanated by it), and thus expressed as a percentage. More precisely, the ratio of transmitted flux to incident flux under specified conditions.

tuning curve Result of an electrophysiological test in which the receptive field properties of neurons are gauged; the maximum sensitivity (or responsiveness) is defined by the peak of the tuning curve.

variable A measurement that can in principle assume any value within some range.

vision The process by which the visual system (eye and brain) uses information conveyed by light to generate appropriate visually-guided responses.

visual angle The angle between two lines that extend from the observer's eye to different points in space.

visual field The area of visual space normally seen by one or both eyes (referred to, respectively, as the monocular and binocular field).

visual perception A manifestation in consciousness of the empirical significance of visual stimuli (and not, therefore, a necessary accompaniment of vision, since vision often occurs without any particular awareness of what is being processed, responded to and thus in some sense being seen).

visual percepts Mental constructs that represent the empirical significance of light stimuli in consciousness, and which allow the observer to reflect upon visual experience.

visual processing Transformations carried out on the information in a retinal stimulus.

visual qualities The descriptors of visual percepts (e.g., brightness, color, depth, form, motion etc.).

visually guided responses An observer's actions undertaken in response to visual stimuli.

wavelength The interval between two wave crests or troughs in any periodic function; for light, the standard way of measuring the energy of different photons (measuring the frequency of photon vibration is another way).

Index

adaptation-level theory, 60
aerial perspective, 11
ambient optical array, 8
apparent distance, 63–65

Bayesian approach, 22
biological significance of image-source
    relationships, 20
biological rationale for empirical ranking, 23

Campbell, Fergus, 100
carpentered environment, 75, 93
computer vision, 9, 99–100, 101
conjoint probability distribution, 104
contrast sensitivity function, 100
cumulative probability, 29, 30, 41–44, 51–60,
    78–79

Delboeuf illusion, 47–48
    comparison of inner circle with a single circle,
        56
    comparison of outer circle with a single circle,
        58–60
    point of perceptual transition, 57–58
    statistical analysis, 54–55
depth processing theory, 95

Ebbinghaus illusion, 47–48
    changing the diameter of surrounding circles,
        49–52
    changing the interval between target and
        surrounding circles, 53–54
ecological optics, 8, 71
edge detectors, 99, 100, 101
egocentric distance, 63

empirical information, 12
empirical ranking, 22–23
end-stopped cells, 105
equidistance tendency, 64–65, 68
Euclidean space, 33, 63, 70
extrastriate visual cortices, 104

feature detection, 9, 99, 102, 104
Fitzpatrick, David, 102
foreshortening, 34
Fourier analysis, 100
Fourier spectrum, 100
Fourier theorem, 100

geometrical illusions, 3–5
Gibson, James, 8, 71
Gillam, Barbara, 7
Gregory, Richard, 6, 83, 84

Helmholtz, Hermann von, 5, 6, 7, 9
Hering, Ewald, 5
Hering illusion, 3, 5, 38, 39, 44–45
heuristics, 6
hierarchical organization of the visual system, 104
higher-order visual areas, 104
historical explanations
    of apparent distance, 70–71
    of apparent line length, 34–35
    of size contrast and assimilation, 60–61
    of the Müller-Lyer illusion, 82–84
    of the Poggendorff illusion, 95–96
Hubel, David, 9, 97, 99

independent component analysis, 21
information redundancy, 21

interpretations of visual neuronal function, 101–103
inverse optics problem, 1, 2

laser range scanning, 15–18
line length, 25–26, 105
    as a function of orientation, 26
luminance contrast boundaries, 26

marginal probability distribution, 104
Marr, David, 9, 99, 100, 101
monocular cues to depth, 11
motion parallax, 11, 71
Müller-Lyer, Franz, 5
Müller-Lyer illusion 3, 5, 6, 7, 73–75, 93
    statistical analysis, 77–79
    statistical analysis of different types of scenes, 80
    statistical analysis of variant stimuli 80, 81

natural scene geometry, 15–17
neuronal receptive field properties, 98
    genesis of, 99–100
    tuning properties, 99
neuronal receptive fields, 21, 97–99
    classical, 98, 105

occlusion, 11
Oppel, Johann Joseph, 5
orientation selectivity, 98, 102

perception of visual space, 10, 11
perception of angles, 37–39, 105
    neural mechanisms of, 45–46
perceived distance
    at eye level, 64–65, 69
    of objects on the ground, 64–65, 69
perspective, 12, 71
physical basis
    of angle perception biases, 41
    of line length biases, 33–34
    of the Ebbinghaus illusion, 52–53
    of the Müller-Lyer illusion, 81, 82
    of the Poggendorff illusion, 93–95
physical sources
    of Müller-Lyer stimuli, 75–77
    of straight lines, 27–29
    of size contrast stimuli, 49
Poggendorff, Johann, 5, 85
Poggendorff illusion, 3, 5, 6, 73, 85–87
    statistical analysis, 87–89
    statistical analysis of additional features, 91–93

statistical analysis of different types of scenes, 93
statistical analysis of the main effect, 89–91
Ponzo, Mario, 5
Ponzo illusion, 3, 5, 7, 8, 55
predicting
    apparent line length, 30–31
    apparent angle subtense, 41–44
    perceived distance, 66–70
    the Müller-Lyer effect, 79
    the Poggendorff effect, 89–93
primary visual cortex, 98, 99, 104
primary visual pathway, 104
probabilistic conception of visual processing, 103–105
probabilistic framework of vision, 10
probability distributions
    of the sources of lines, 27–31
    of the sources of angles, 39–41
    of physical distances, 65–66

representation of retinal images, 9, 103, 104
retinal images and the physical world, 19
range image database, 15–19
    limitations of, 17–18
Riemann space, 70
Robson, John, 100
rule-based scheme of vision, 9

specific distance tendency, 64, 68
size contrast stimuli, 47–49
size assimilation stimuli, 47–49
sparse representation, 21
spatial frequency, 100–101, 105
spatio-temporal filtering, 101, 102
statistical properties of natural images, 20–21

T-illusion, 3, 5, 26
texture gradient, 12, 71
Thiéry, Armand, 6
tilt illusion, 39, 44–45

unconscious inferences, 5

vertical-horizontal illusion, 3, 5

Wiesel, Torsten, 9, 97, 99
Wundt, Wilhelm, 37

Yarbus, Alfred, 18, 27

zero-crossings, 35
Zöllner illusion, 38, 39, 44–45